STEAM LOCOMOTIVE CASUALTY REPORTS

STEPHEN MOURTON

RUNPAST
PUBLISHING

Contents

Introduction   5

Chapter 1
Locomotive Casualty Reports   6

Chapter 2
Reports for 1958 and 1959   7

Chapter 3
Reports for 1960   11

Chapter 4
Reports for 1961   32

Chapter 5
Reports for 1962 and 1963   55

Chapter 6
Barnwood Locomotive Depot   66

Appendix 1
Locomotive mileages shown in the Casualty Reports   72

Appendix 2
LMSR Locomotive Casualty Report System   74

Appendix 3
Common causes for engine failures   78

Copyright ©2005 Stephen Mourton

ISBN 1 870754 62 X

Runpast Publishing
10 Kingscote Grove, Cheltenham, Glos GL51 6JX

Typesetting and reproduction by Viners Wood Associates
Printed in England by The Amadeus Press Ltd., Cleckheaton

Acknowledgements

Thanks are due to the following individuals who have provided the valuable information and insights that made this book possible: Roy Smith for originally bringing the Barnwood Locomotive Casualty Reports to my attention; Pat Cook who was in the Control Office at Gloucester Eastgate and Northgate Mansions, Gloucester for a number of years in the 1950s and 1960s; Roy Denison who worked in the District Running & Maintenance section of the Trains Office at Northgate Mansions between 1960 and 1963; and Mike Randall, long-serving railwayman and a fireman at Barnwood depot from the late 1950s until its closure. Also thanks to another Barnwood fireman, Derek Smith, for reading the manuscript and adding some of his own footplate experiences. Robin Stanton's photos, taken from Hatherley signal box and elsewhere, have proved very useful for this book. Finally, it has been a great pleasure to meet and converse with Fred Cole, shed master at Barnwood from March 1958 to April 1964, who kindly provided information on a number of matters.

Selected bibliography

LMSR Locomotive Casualty Report System – methods used for reporting, and dealing with, engine failures
Harold Rudgard, The Railway Gazette, 1944

Motive Power Organisation and Practice
Harold Rudgard, LMS Railway, 1946

Steam Locomotives in Traffic
E A Phillipson, Locomotive Publishing Co, 1949

A Breath of Steam *W G F Thorley, Ian Allan, 1975*
This book includes a table of loco failures at Bedford in 1932 showing: engine number; owning depot; nature of casualty; cause of casualty, allowing some comparison with the situation at Gloucester in the early 1960s.

Living with London Midland Locomotives
A J Powell, Ian Allan, 1977

The Stanier 4-6-0s of the LMS
J W P Rowledge and Brian Reed, David & Charles, 1977

Organised Chaos! – Bath Green Park Running Foreman's Log Book 1960/61, *Somerset & Dorset Railway Trust, 1989*

Trains Illustrated magazine for September 1953 includes a brief article looking at shed procedures – **On Shed** by Alan Jeffryes

British Railways Magazine – Western Region, June 1962 had a picture feature – **Jobs to be Done** – at Gloucester Barnwood Depot.

Esso tank trains to Bromford Bridge, Birmingham from the Fawley refinery near Southampton and from Avonmouth Docks, Bristol started running on the Birmingham-Bristol line in 1960 with BR 9Fs being the usual power. These were heavy trains so needed BR's most modern and powerful steam locos to perform well if the trains were to maintain their schedules. Unfortunately this did not always happen and 92248 which is thundering up the Lickey incline on a working from Avonmouth was a failure in several casualty reports. This view is dated 13 April 1964 and there were three banking locos — 0-6-0PT 8400 and 8418 plus 9F 92230.

Derek Cross

4

INTRODUCTION

The newly nationalised British Railways started operations on 1 January 1948 and something at which it excelled, like all large organisations, was churning out paperwork – stacks and stacks of 'bumf'. Whilst some has survived, a lot more got destroyed, understandable bearing in mind the quantities produced. Upon impending closure one loco shed filled an engine pit with paperwork, all viewed as mundane stuff at the time, not worth saving, and set it alight. No doubt much historically interesting material met a similar fate, but sometimes discarded documents were rescued by enthusiasts from the shattered, vandalised remains of what used to be orderly workplaces, then got stored and forgotten in attics and basements. Over four hundred *Locomotive Casualty Reports* dated between 1958 and 1963, salvaged from Gloucester Barnwood motive power depot, came to light recently. They are analysed and amplified in this work, providing a valuable insight into BR operations during the last years of steam and the changeover to diesel power; and help to record for posterity some of those eventful days. Many locos on a variety of workings and from a multitude of sheds on the London Midland, Western, Southern and Eastern Regions are mentioned in these reports.

A lot of failures recorded in the casualty reports tend to spring from lack of routine servicing and maintenance. In the early 1960s, there were still school leavers wanting to work on the footplate but recruitment for such jobs as fire dropping, ashing out, tube sweeping, coaling up – the *really* dirty stuff – was more difficult. Barnwood had a disposal crew on at least two shifts – usually a restricted driver and a passed fireman – who attended to seven locos per shift. In addition there was a preparation crew for making sure a loco was ready to leave shed on time.

Many railway personnel were lured away by the attraction of jobs in shiny new factories with regular hours, better pay and prospects. Skilled fitters were in demand throughout industry, so BR had to fight to recruit and keep good people. On the other hand, there were still a goodly number of personnel at Barnwood who had started work on the Midland Railway and had over 40 years service to their credit.

The future of British Railways was under critical scrutiny – with closures, rationalisation and modernisation reducing staff numbers, career prospects were uncertain if you joined the railway. It all added up to a struggle in just keeping the services going – though railwaymen and women at all levels still had a strong sense of duty. And it was a period of enormous changes in motive power, with unfamiliar steam types increasingly appearing – some of which are mentioned in the narrative – followed by the wholesale replacement of steam by diesel power.

Traffic on the main line through Gloucester came from the Midlands and North to the West of England, South Wales and the Somerset & Dorset line. Expresses included the *'Devonian'* from Bradford to Paignton and Kingswear and *'The Pines Express'* from Manchester to Bournemouth, reached via Bath and the Somerset & Dorset line, which also had through coaches from other places such as Liverpool. There were several trains from Newcastle, York and Sheffield, including important mail trains, plus workings from Derby and Nottingham. The expresses were not noted for their speed, often being quite heavy trains. In the summer, there were many extra passenger trains to cater for holiday traffic, especially to popular West Country resorts and Bournemouth. Freight of all sorts was also very heavy throughout the year. Coal and steel moved in both directions, mainly up from South Wales and down from the Midlands. Gloucester was not a major loco changing point for through passenger trains to and from the Midlands and north – the South Wales to Newcastle express was a notable exception, swapping a GW for an LMS engine at Central station and vice-versa.

This study tends to use the familiar *Ian Allan ABC* descriptions for locos, such as 3F, 4F, 'Jubilee'. But Barnwood and other Midland men would invariably call these types 'Three Freight', 'Big Goods', '5X'. And they would not say '44123' but 'Forty-one Twenty-three'.

Some details of Gloucester Barnwood's engine allocation and workings are given later in the book. The majority of the depot's work was centred on the former Midland Railway Birmingham-Bristol/Bath main line and associated branches. Other depots also provided engines and crews for workings through Gloucester, prominent amongst them being Saltley, Bristol Barrow Road, Bath Green Park, Derby, Sheffield Millhouses and Leeds Holbeck. But 'foreign' engines appeared every day from other depots, adding variety to train observations.

During the period covered by this work, there were organisational changes on BR. The Western Region's Commercial, Operating and Running & Maintenance Departments combined early in 1960 to form the Traffic Organisation, which had four geographical divisions, sub-divided into various districts including Gloucester. The Midlands Division based at Birmingham had jurisdiction over Gloucester, whose District Offices moved from Worcester to Northgate Mansions, Gloucester, in September 1960. Jack Powell was District Traffic Superintendent; reporting to him were District Running & Maintenance Officer George Bartlett and his assistant Guy Kerry, to whom Barnwood shed reported. But many of the casualty reports still refer to the previous title of 'DMPS' – District Motive Power Superintendent – so this book has standardised on that nomenclature for the sake of continuity.

CHAPTER ONE
LOCOMOTIVE CASUALTY REPORTS

The London Midland & Scottish Railway *Locomotive Casualty Report System* was one of the standard procedures of its motive power department and was subsequently adopted by BR. It detailed the methods used for reporting, and dealing with, engine failures. Extracts from the document issued to running sheds explaining the system and dated 1944 is contained in Appendix 2 of this book. The main instigator was Harold Rudgard, Superintendent of Motive Power for the LMSR, who wrote in 1946: 'The casualty report system now in use is an extension of the ideas incorporated in the original system of reporting casualties on the old Midland Railway and I was closely associated with the late Sir Cecil Paget in bringing into being this method.'

A report had to be completed in the following circumstances:
1. If during the working of a train any defect developed in the engine or tender whereby efficiency is impaired and three minutes or more time is therefore lost (later altered to five minutes for passenger trains and ten for freight).
2. The failure of an engine, after being fired, to work its train to within three minutes of time (later altered to five minutes for passenger trains and ten for freight) through any defect which should have been found by the examiner or driver or was due to a concealed defect.
3. The fusing of lead plugs irrespective of whether or not time was lost.
4. Time lost due to shortage of steam, inferior coal, slipping, mismanagement by enginemen, vacuum brake irregularities, shortage of water and problems with steam heating.

By way of comparison, a failure on the GWR was deemed to have occurred:
1. *Where a mechanical or boiler defect in an engine results in delay to a train.*
2. *Where an engine, owing to any mechanical or boiler defect, has to come off its train, or turn, short of destination.*
3. *Where any failure as above does not cause delay, but necessitates an engine being kept out of traffic to such an extent as to render it unable to work its next ordinary booked turn.*
And on the Southern Railway:
1. *An engine which has to come off its train owing to any mechanical or boiler defect, even if no delay is caused, is to be considered a mechanical failure.*
2. *An engine which through a mechanical or boiler defect causes a delay to its train of 10 minutes or over is also considered to be a mechanical failure.*

The reports were accompanied by relevant correspondence such as the driver's explanation of reasons for the time lost, details of the loco's servicing history from its home depot, and the shed master or foreman's comments from the depot which received the failed engine.

The reports analysed in this book dealt with locomotive casualty reports which:
1. Concerned an engine crewed by Barnwood men.
2. Related to one of Barnwood's own engines. No matter where it failed, a report found its way to Barnwood.
3. Covered an engine whose previous servicing before failure had been at Barnwood.
4. Concerned failure by an engine whilst working a train in the Gloucester area on the Birmingham-Bristol/Bath ex-Midland Railway route and associated branches. The loco usually, but not inevitably, went to Barnwood depot for scrutiny by a foreman and possible remedial attention.

Copies of reports were distributed as appropriate by the initiating shed to:
> The loco's home depot
> The loco crew's home depot
> District Motive Power Superintendent
> The Chief Mechanical Engineer at Derby or Crewe for mechanical failures
> District Traffic Superintendent
> Divisional Traffic Manager
> Assistant General Manager, Traffic, Paddington

It should be emphasised that the ex-GW steam loco depot at Gloucester Horton Road, code 85B, usually dealt with failures on other routes in the area and forms no part of this study. However GW loco casualties were often found at Barnwood receiving attention after getting a hot box; they came not just from Gloucester but places such as Swindon, Worcester, Bristol and South Wales and had been doing so since 1936 or earlier, according to the *Railway Observer* magazine.

There are a handful of casualty reports from the late 1950s and lots from 1960, 1961 and 1963, but those for 1962 have disappeared, with two exceptions. Judging from records in the author's possession for locos repaired at Barnwood in 1960 and 1961 – which have been listed in Chapter 6 – there were other failures in traffic, certainly of Barnwood's own locos, for which no corresponding casualty report has been seen. The shed office would have prepared summary sheets or books listing every report; if these survived is not known. However the amount of available material – well over four hundred reports as already stated – gives a very good idea of the problems faced by motive power depots and enginemen in keeping services operating during the late steam era. Not all the reports available for this study have been the subject of comment in the narrative; some minor ones have been listed in summary tables to avoid repetition.

Details of common causes for engine failures are shown in Appendix 3.

CHAPTER TWO

REPORTS FOR 1958 AND 1959

'...the right hand driving crank pin breaking, causing extensive damage to the coupling rods and putting all engine wheels out of gauge.'

The oldest report in this survey is the only one surviving from 1958 and is also the most drawn-out.

On Saturday 3 May the 1.35am class 'E' freight from Birmingham Exchange Sidings to Westerleigh, load 29, set off with Saltley's 4F 44230. There were no reported problems until just after leaving Gloucester Eastgate station, with a Bristol Barrow Road crew on the engine. The loco came to a sudden stop at California Crossing, just a few hundred yards away, 'due to the right hand driving crank pin breaking, causing extensive damage to the coupling rods and putting all engine wheels out of gauge' according to the Barnwood shed master's report. This blocked the main line, sometime after 4am if the train was running to time at that stage. Getting the engine moved to Barnwood shed would have presented problems. It was examined at the depot by an inspector from the LMR Chief Mechanical Engineer's department and put into store in Barnwood shed yard without any repairs being done. And there it remained, a familiar sight for many, many months. It was shown in the LMR locomotive stock book alteration list number 471 for the week ending 12 December 1959 as being withdrawn for breaking up. But it was still at Barnwood on 1 September 1960 when the shed master wrote to his boss at Worcester, the Divisional Motive Power Superintendent, asking how it was proposed to dispose of the engine 'bearing in mind that the wheels will have to be changed before it can be moved on its own wheels.'

The reply dated 11 October 1960 stated that a set of temporary wheels for the loco had been sent from Derby Works on 5 October. The old wheels were to be returned to Derby, which was done on 9 November. Eventually the loco was hauled away on 28 November to meet its fate, over two and a half years after failing at Gloucester.

Prior to 44230's long stay, it is believed the record for the longest period spent at Barnwood after failure was held by 42863 of 56A Wakefield. It failed on 14 July 1957 and did not leave until 29 October that year.

44230 looks very woebegone – not surprising as it spent over two and a half years out of use in Barnwood shed yard after failing at Gloucester in May 1958; it never steamed again and was towed away in November 1960, being scrapped at Derby Works in February 1961.   *Brian Miller Collection*

LOCOMOTIVE CASUALTY REPORT (MECHANICAL)

Barrow Road ....Depot ...... Bristol .....District Engine No. 44230 ......Class. 4F

Allocated to.... Saltley ....Depot. Driver.... R. Ridout ....82E.....(No.....) T. Godwin ....Fireman.....82E....(No......)

working the 1.35a ...m. (class) "F" ....train from.... Exchange Sdgs ....to.... Westerleigh

on.. Satur ..day, the.. 3rd ..day of.. May ....19..58.. Assisting Engine No................Class................
Assisted by

Allocated to....................Depot. Driver...................(No.........) Fireman....................(No.........)

became a casualty at.... Gloucester ....causing a delay of.. not known ..mins. Engine changed at............................

No. and Class of engine
working forward ....Load of train.... 29 ....for engine. Regulation load 29

NATURE OF CASUALTY

R.H. Driving Crank Pin broke and Coupling Rods bent (came to a stand
at Gloucester - California Crossing after starting away from Gloucester).
Bristol Driver relieved and engine and train left on Main Line.

| YOUR REF. | | 17 | | OUR REF. | RM K2818 | B.R. 14303/71 |
| DATED | | BRITISH TRANSPORT COMMISSION | | DATE | 11/10/60 | |

BRITISH RAILWAYS

DISTRICT TRAFFIC SUPERINTENDENT'S OFFICE

TO Mr. F.A. Cole,
Gloucester Barnwood.

RUNNING & MAINTENANCE
85E FROM
12 OCT 1960
W.R. Extn.... 74
GLOUCESTER (BARNWOOD)

WESTERN REGION
GLOUCESTER STATION

Engine No. 44230/21A - failed at Gloucester 3/5/58.

With reference to your letter of the 1st ultimo, please note that a
set of temporary wheels are being forwarded to Gloucester Barnwood for fitting
to the above mentioned locomotive to enable it to be forwarded to Derby Works
for breaking up.

The wheels have been loaded in Wagon No. M.72883 and were despatched
on the 5th instant. Please arrange to return the old wheels taken from
locomotive 44230 in this wagon to the Works Manager, Derby, advising me
when despatched, also letting me know the date the engine is despatched.

For J. POWELL, DISTRICT
RUNNING AND
MAINTENANCE OFFICER

REPORTS FOR 1959

'Delay caused to train through inferior class of engine'

44688 of 17A Derby shed had the 9.35am Bradford-Bristol express, M251, from Leeds on Sunday 26 April, but gave up short of steam at Gloucester. Apart from a dirty fire and poor quality coal, the left piston gland was blowing and the trailing boxes were knocking badly. The driver, as was usually the case with engines doing badly, had presumably wired ahead for a fresh engine because there was no delay in changing over locos – but the train was already 41 minutes late arriving in Gloucester.

A Barnwood driver was not impressed with his Saltley 4F 43911 on Saturday 9 May, when he had the engine on the 1.32pm Derby-Bristol class 'C' parcels loading to 365 tons. Although losing only nine minutes, he had expected rather more powerful motive power for the train as it was diagrammed for a Holbeck class 5 loco. 'Delay caused to train through inferior class of engine' was his first report, while an amended one added 'and superheater elements blowing. Repairs booked at Bristol Loco.' So, in the circumstances, the performance was not too bad.

A more modern Class 4, BR Standard 75021, a Bristol Barrow Road engine, failed at Gloucester on Friday 14 August due to a problem with the vacuum relief valve, just before working the 11.30am stopping passenger to New Street. Delay due to this was 17 minutes and the new engine for the working was BR Class 5 73016 of Millhouses shed, Sheffield, so there was more powerful motive power on this occasion.

The '75' Standards had replaced most – but not all – of the Midland '2P' and Compound 4-4-0s at Bristol Barrow Road and Gloucester Barnwood sheds within the last twelve months. But the old locos refused to die. Barnwood's last compound, 41123, was not just station pilot at Eastgate during August, but got as far as York on a class 'A' passenger from the Sheffield direction. One of Barnwood's 2Ps, 40489, was also gainfully employed, being noted piloting 73046 on the 8.30am Cardiff-Newcastle on 15 August, while Derby's Compound 40925 doubleheaded with 45648 on the down Newcastle-Cardiff into Gloucester that day.

The latter combination was normal for that express, so something somewhere had gone awry on Monday 24 August when the train was seen at Cheltenham with 9F 92158 of 18A Toton.

A very unusual type for the route seen by the author on Sunday 23 August was J39 64930 on the 4.55pm Bristol Temple Meads-Birmingham New Street stopper. This loco, from 9G Gorton depot, had apparently arrived in Bristol on an overnight freight. Another J39, 64789, worked a special freight into the area in October 1959.

One of Barrow Road's BR Class 5s, 73031, was in trouble on Saturday 7 November while in charge of the 7.40am express from Bristol to Bradford, loading to 333 tons. Its left hand piston rings had broken, so it came off the train at Gloucester and returned light to Bristol for attention. The report notes that the engine had covered 17,000 miles since the rings were replaced in June 1959.

73031 spent time at Rugby Locomotive Testing Station, along with 73030, and was equipped from new with a Westinghouse air pump, used on trials with fitted coal trains. One test train was on 24 January 1954 'consisting of 73030 and 73031 hauling a recording car, 50 Westinghouse fitted coal trucks and a Westinghouse fitted brake van' according to the Railway Observer. 73031 also completed 69 runs at the Rugby plant undergoing superheater and cylinder efficiency tests for several months in 1958, the last 4-6-0 to be tested there.

The replacement loco for the train is of interest, being Eastern Region 'B1' 61370. Earlier in the year 61370 had been allocated to 30E, Colchester, but, displaced by modernisation, it was transferred to Sheffield and, along with other members of the class, became a regular in the remaining steam years on the Birmingham-Bristol main line. And there was another 'B1' in the area that day, 61083.

A few days later, 16 November, saw a much rarer Eastern Region visitor when 'V2' 60954 worked through to Bristol on the 12.48pm express from York, much to the consternation of Western Region management, who sent it back light engine the same day.

To add to the unusual classes, 'K3' 61961 worked a southbound car train through Gloucester on 21 November, while 'B1' 61152 was also seen that day. Barnwood men worked another 'V2' – 60839 – north from Bristol in the middle of December.

The 7.25pm Bristol-Newcastle was an important train, taking the mails north. Millhouses 73048 was at the head of the 358 ton load on 21 December with Barnwood Driver Bill Dipple and Fireman F G Hooper. 21 minutes were lost running to Birmingham due to shortage of steam, with a dirty fire and ashpan; the blower pipe was also found to be fractured when the engine was inspected at Saltley shed. A fresh engine took the train north from New Street.

After repair at Derby Works one of Barnwood's own, 4F 44296, was working home on 22 December with the 8.55pm Derby Chaddesden Road to Birmingham class 'C' fitted freight. It had only done 700 miles since being overhauled, but experienced problems with the small ejector, being unable to create the required amount of vacuum, so it worked the train loose coupled.

A couple of 'WD' 2-8-0s, Nos. 90565 and 90685, transferred to Barnwood in November 1959, but were 'never worked by Barnwood men and only come to the depot for washouts etc.' according to the shed master, when he dealt with 90685's serious failure at Woodford Halse on 24 December, the right cylinder casting being broken. 90685 left Gloucester the previous day with 85B Horton Road men. Cause of the failure was that the front cylinder cock was blocked. The loco's 'X' examination at Barnwood a couple of days earlier had not revealed any problem and the shed master stated 'I am quite satisfied that no member of my mechanical staff is responsible for blanking the cock off.' The shed master at Woodford agreed that no-one was at fault for this problem, and his boss, the DMPS at Rugby, concurred. 90685 was proposed for a visit to shops; it had covered 33,000 miles since overhaul at Crewe in June 1958.

Woodford Halse depot, 2F, on the Great Central main line between London and Leicester, had a large allocation of 'WD' 2-8-0s, so was very familiar with this type. Its 'WD's worked down to Rogerstone, near Newport (Mon.) via the Stratford-on-Avon & Midland Junction route and Ashchurch during this period. Gloucester men took over from Woodford Halse crews at either Broom Junction or Ashchurch. According to Mike Randall, Barnwood men did work their home-based 'WD's from time to time, though this type's allocation to Barnwood was short-lived in the early 1960s.

'Jubilee' 45579 *Punjab* was 40 minutes late leaving Bristol Temple Meads with the heavy 11.45pm parcels train to Derby on 30 December. The Gloucester driver said the delay was due to the loco's injectors failing and complained that it had not been properly prepared for the job at Barrow Road shed.

Right: One casualty report has 61370 of 41A Sheffield Darnall taking over a Bristol-Bradford express on 7 November 1959 at Gloucester from a sick 73031. A few months earlier, 61370 was shedded at 30E Colchester and the chances of it appearing at Gloucester would have been very slim indeed. In this picture, 61370 is seen near Churchdown on a northbound relief passenger. 'B1's only became a regular sight in the last few years of steam due to boundary changes between the Midland and Eastern Regions, plus modernisation in East Anglia shifting more of the class to the Sheffield/Rotherham area. Sometimes they proved tricky for crews unused to their types of injectors, leading to problems.

Below: Locos based on the Somerset & Dorset line from Bath to Bournemouth were often seen at Barnwood shed for attention to hot boxes. Midland 3F 0-6-0s frequently had this problem and 71H Templecombe's 43194, working home from overhaul at Derby Works, had its centre drivers sent back there for repair after getting a hot box. Interesting casualties attended to at Gloucester from other sheds on the Birmingham-Bristol and Bath lines included 538XX 2-8-0s, Sentinel 4-wheel shunters 47190 and 47191 and the famous Lickey banker, Fowler 0-10-0 58100. *Brian Miller collection*

CHAPTER THREE

REPORTS FOR 1960

The year 1960 was 'the calm before the storm' as far as the railway in Gloucester and its motive power was concerned. Passenger and freight turns on the ex-Midland Birmingham-Bristol line were still steam worked, with just the occasional appearance of diesel multiple units on excursions from the Midlands and perhaps a diesel shunter lurking in a goods yard. Top express motive power were the 6P/5Fs, mainly 'Jubilees' plus a few 'Patriots', operating from their traditional sheds, such as Bristol Barrow Road, Sheffield Millhouses and Leeds Holbeck. Associated branches in the area like Ashchurch to Upton-on-Severn and Coaley to Dursley were still in business and steam-worked.

JANUARY TO MARCH 1960

'...have special attention given to the servicing and coaling of the engine working this important service.'

Sometimes a fresh engine was not much better than the casualty it replaced. On 4 January a Saltley crew had problems maintaining boiler pressure on Barrow Road's famous record-breaking 'Jubilee' 45660 Rooke while working the 12.48pm York-Bristol express. So it came off at Gloucester in favour of classmate 45649 Hawkins, which was taken forward by a Barnwood crew. The driver reported 'Before I left I had engine with full boiler pressure and boiler of water, I considered I could get to Bristol with this engine but had to stop at Berkeley Road to recover boiler and pressure, causing a delay of about 8-10 minutes. After leaving Berkeley Road had no further trouble with engine.'

In September 1959, 45660 had been involved in a collision at Barrow Road depot coaling plant, in which a driver was injured and 'Pug' 0-4-0ST 51221 damaged. On 17 December 1959 51221 was despatched to Horwich Works for scrap.

While a number of older Midland loco classes had disappeared from the area by 1960, one type still performing hard work was the 3F 0-6-0. These free-steaming engines could suffer, like some Midland types, from hot boxes and this is what happened with Saltley based 43242 on 9 January. It left a Washwood Heath to Westerleigh goods at Gloucester with the right driving and left trailing axles hot. 43242 was positioned on the wheel drop at Barnwood, the driving and trailing wheels and boxes being removed and sent to Derby Works for repair; the loco was ready for service again on 15 February. Another member of the class with a similar problem in this period was 43484 which worked the early morning Barnwood to Bromsgrove pick-up freight on 5 February, returning light engine as booked, but found to have a hot box on inspection at the shed. 43484's damaged boxes were reconditioned and the journals turned.

Some Midland men claim that the relatively high incidence of hot boxes on their 0-6-0s was due to the fact that they regularly sustained speeds for which they were not designed. What Western men would say about that is probably succinct but unprintable!

The 8.30am Cardiff-Newcastle express was a heavy train, requiring two engines from Gloucester Central station, where the Midland took over from the Western Region, if the load was over 385 tons. When just one loco was rostered for the working, it needed to be in 'good nick'. Unfortunately this appears not to have been the case on 13 January with Millhouses 'Jubilee' 45683 Hogue, which is recalled as usually being in totally filthy external condition. The actual load was 390 tons, a tad over the booked limit. The train arrived at Gloucester 16 minutes late, but lost another 30 minutes in running to New Street. The Barnwood driver claimed the loco was short of steam and the lateness was made worse by permanent way and signal checks. The loco got as far as Derby where it was replaced – by 44818 – and examination revealed a number of element joints leaking due to wastage. It must have had a quick repair – or more likely was just put back into traffic – as it was recorded in the casualty report's follow-up as working the 10.55pm from Leeds to Sheffield later that day. However the DMPS at Worcester was concerned about the performance of the loco and stated 'I am asking the Gloucester Barnwood shed master to have special attention given to the servicing and coaling of the engine working this important service, with special reference to the tubes and tubeplate.'

A fairly rare 4F for the line was 44444 of Birmingham's Monument Lane depot. Locos from the 'North Western' side of Birmingham did not have any regular turns on the Bristol line at this time. It had a fairly routine problem, unable to create vacuum when coupled up in Bristol Temple Meads station to the 6.30pm stopper to New Street on 13 January. But Barnwood Driver W Lane showed initiative by effecting a repair rather than calling for a new engine and the train left 24 minutes late, with a few minutes regained on the journey to Gloucester.

The up 'Devonian' on 14 January had a rather protracted journey. Barrow Road's 45572 Eire was unable to heat the train properly, so came off at Gloucester, causing a 32 minute delay. Its replacement, Standard 5MT 73016, was apparently not steaming freely, but made it to Derby, losing a further 33 minutes, where 44658 took over and ran through to Leeds.

A couple of days later, on a frosty 16 January, another 'Jubilee' was in trouble, this time 45610 Ghana of Derby which had charge of the 12.43pm Newcastle-Bristol express. It got to Standish Junction seven miles south of Gloucester at 9.0pm before giving up due to poor steaming. It was scheduled to pass Standish at 7.36pm, so was pretty late – of course not all the lateness was necessarily down to the engine but perhaps the Bristol driver might have been better off getting a fresh loco at Gloucester Eastgate station. On examination of Ghana at

Above: 'Jubilee' 45683 Hogue *was allocated at 41C Sheffield Millhouses which had a number of turns on the Birmingham-Bristol line and whose locos feature quite often in the casualty reports. 45683 had problems on the Cardiff-Newcastle express on 13 January 1960, being short of steam. In this picture at Hatherley Junction, the loco, in its usual filthy state, is heading a northbound express.*
R Stanton

Barnwood shed it was found that about 12 tubes in the bottom row were blocked, the brick arch was in dirty condition and there was a hole in the smokebox. Derby shed's report showed the tubes had last been cleaned on 7 January and there were no previous reports of bad steaming.

Another Derby 'Jubilee' with problems was 45627 *Sierra Leone*. It was working the 7.35am Nottingham-Bristol passenger on Saturday 30 January – a train which had been the regular working of Derby's sole 'Patriot' 45509 *The Derbyshire Yeomanry* until it was transferred away in the summer of 1958. 45627 lost about 10 minutes between Gloucester and Bristol with the Barnwood driver reporting: 'Engine not steaming and boiler priming. We had to stop overtime at Berkeley Road station to fill boiler and get more steam. The Derby driver that I relieved at Gloucester told me that he had trouble all week with the same engine and Saturday was the worst of all. It appeared to me that nothing had been done to the engine.' However Derby shed claimed there were no reports of any problems with 45627.

The 8.30am Cardiff-Newcastle express was experiencing poor loco performance again on 8 February with 45594 *Bhopal* losing 24 minutes in running between Cheltenham and Bromsgrove. No doubt mindful of the DMPS's recently expressed concern about the preparation of locos on this train, the shed master sent a covering letter with the Barnwood driver's report. The driver said he had asked for the loco's tubes to be cleaned before leaving shed, but the tube sweeper had not

signed on for duty, so nothing was done. The shed master confirmed this, but said that when the loco worked to Gloucester the previous day on the 12.15pm from York, that driver had reported 'no repairs' and therefore no attention had been given to the tubes by the night shift.

A shed not often represented at Gloucester was Coalville, 15D, but its 4F 44260 was removed from a Washwood Heath-Westerleigh freight at Tramway Junction on a snowy 11 February, after the right hand big end cotter went missing. A 135 minute delay ensued before 43975, Saltley-based from January 1960, but coincidentally a Coalville engine before that, worked the train forward.

Barnwood men did not have a good journey on the Sundays 6.28pm Newcastle-Bristol class 'A' passenger of 14 February, which they worked from Derby. The loco, 58 minutes late off Derby shed, was 'Black 5' 45299 – of Stoke depot, 5D, according to the report but actually from Crewe South, 5B, again not a shed which normally provided engines for this turn – so maybe it was an emergency replacement. The driver reported time lost between Burton and Kingsbury Junction, where a stop was made to fill the boiler, as well as permanent way and signal checks. The scheduled time from Derby to New Street was 52 minutes, but it took 1 hour 20 minutes, with 18 minutes of the loss put down to the engine. 45299 was replaced at Birmingham and Driver F E Ford was very dismissive: 'Engine not fit to work passenger trains.'

The 4.48pm class 'C' express fitted freight from Bristol St Philip's to Leeds Hunslet Lane was one of the most important trains on the line – known unofficially by some staff as the 'Million Dollar Flyer' due to carrying valuable cargo – so any delays were a cause for concern. This was often worked by a 'Crab' as it was on 16 February, with a Nottingham crew aboard 42900 hauling 34 loaded wagons. The loco succumbed at Gloucester when the exhaust injector failed and the live steam injector kept blowing out. Barnwood could not match the Crab's class 5 power, providing a 4F as replacement, its own 43924, but at least the train was underway after just a 20 minute delay. Examination of 42900 found small coal lodged in the exhaust injector, while the water regulating valve was loose on the live steam injector.

Millhouses 'Jubilee' Bhopal had problems again on 21 February whilst working the 7.5pm Newcastle-Bristol mail train. The engine was priming badly and Driver F E Ford reported stopping at Cleeve to fill the boiler. Eastgate pilot 75009 took the train on to Bristol from Gloucester at 6.0am; it was actually due away at 4.34am. Barnwood was apparently none too pleased to see Bhopal again and sent it back light engine to its home depot for attention. While some drivers managed with one injector, by all accounts Driver Ford definitely needed both injectors to be in full working order.

There is an interesting comparison possible between this loco's performance in the late BR steam age and in the immediate post-war era. On 1 November 1945, Bhopal made an excellent run with Barnwood Driver Cyril Cook at the regulator on the 5.10pm Derby-Bristol. Despite signal checks being encountered throughout the journey, a 40 minutes late departure from Derby had been transformed into an arrival at Cheltenham just 16 minutes down in an overall time of 2 hours 6 minutes. In 1960/61 this same train was allowed 2 hours 8 minutes from Derby to Cheltenham, stopping only at Burton and Birmingham – in 1945 it also called at Tamworth and Ashchurch.

On 22 February, Redditch men were working the 8.27am Birmingham-Evesham-Ashchurch passenger with Ivatt 2-6-0 43047. Somewhere *en route* the right trailing brake block became detached from its hanger and the right trailing side rod got bent. 43047 continued to Ashchurch where Saltley shedmate 4F 44213 took the return working. The damaged items were sent to Derby for straightening.

23 and 24 February saw Gloucester Control at Northgate Mansions kept busy dealing with six engine failures. First was 73136, of 17C Rowsley, with the 6.12am Derby-Bristol express on 23 February. This limited load train was quite fast – averaging 60mph start-to-stop between Bromsgrove and Cheltenham. The loco's problem was caused by the right front valve sticking up, apparently cured by oiling the spindle valves on this Caprotti valve gear engine. 'Jubilee' 45570 New Zealand worked the train from Gloucester. After attention 73136 returned north on the 1.44pm Gloucester to Birmingham stopping passenger.

An earlier stopper that day, the 9.15am Bristol-Birmingham, changed engines at Eastgate as 42825 could not heat the train and 'Black 5' 44963 took over.

Driver F E Ford and Fireman T Westwood did well to keep

to schedule on the 'Pines Express' 10.15am Manchester-Bournemouth, as the driver reported the engine, Derby 'Jubilee' 45598 Basutoland, was not steaming because of very bad coal, also problems due to defective vacuum. As a result, the Eastgate pilot, once again 75009, was called upon to run the express to Bath.

Things then went quiet until near midnight. Westhouses 9F 92113, with Bath Green Park footplatemen, had charge of the 8.25pm Templecombe (10.15pm ex Bath) -Derby class 'C' perishables, due Eastgate at 11.17pm. 92113 could not work further due to a defective injector; it appears the driver had complained about the injector steam valve blowing through, filling the cab full of steam, before the loco left Bath. 92113's replacement at Gloucester was, perhaps by design, a Bath engine, 73052, which took the train northwards. As far as the Somerset & Dorset line was concerned, this train, conveying premium traffic and Burton beer empties from Exmouth Junction and the west, was of equal importance to the 'Pines Express'. 73052 could not have worked through to Derby, as it returned on the 2.0am Derby-Bristol parcels, due Gloucester at 6.47am. This parcels was routed via Stourbridge Junction so did not go through Birmingham New Street.

Into the early hours of 24 February and 8F 48672 of Toton depot arrived at Eastgate station on the 8.15pm class 'J' loose coupled freight from Washwood Heath to Westerleigh and, as Barnwood Driver E Collins tells it, 'Upon making brake application to stop at Barton Street home signal, the steam brake intermediate flexible hose became fractured and was afterwards found broken. Train brought to rest by handbrake and reversing gear.' At least Barton Street crossing gates, just beyond the signal, escaped demolition! After an 80 minute delay 4F 44465 headed the train to its destination. There was an occasion 25 years earlier, in February 1935, when brand-new 'Jubilee' 5594 Bhopal passed a signal at danger and did actually damage the Barton Street crossing gates. It must have been one of the earliest appearances of a 'Jubilee' at Gloucester. The Bristol driver got a one day suspension for this misdemeanour.

A tragedy occurred in the 1950s at Buxton when the steam brake pipe on the footplate of an 8F fractured and the freight train it was hauling ran away out of control. The driver, John Axon, who stayed on the footplate, was killed in the ensuing crash in Chapel station and was posthumously awarded the George Cross.

To round off a busy 24 hours, Saltley 4F 43940 ran hot on the 1.50am class 'J' freight from Washwood Heath to Ashchurch. This was a bit unfortunate, as it had only covered 400 miles since being repaired at Barnwood, presumably for a similar problem, and put back into traffic on 10 February. Now it managed to get to Barnwood with another hot box, after running light from Ashchurch. 44171 took over its northbound return working.

On 26 February, it was once again 'Jubilees' from Derby depot giving Control headaches. 45649 Hawkins failed at Gloucester on the 8.40am Bristol-Sheffield express with injector problems, being replaced by classmate 45607 Fiji with a minimal delay, just four minutes.

Derby's 45648 Wemyss had charge of the down 'Pines

Express' that day, arriving at Gloucester around 15 minutes late, on the face of it not too bad, but a fresh loco was required – this was the second time in four days that the down *'Pines'* had engine trouble. As Barnwood Driver L G E Smith put it, 'this engine in hopeless condition. Brickarch – one row of bricks only. Tubeplate and tubes blocked completely. Engine priming very badly, unable to keep water in boiler. Primed from Birmingham-Gloucester continuously.' He also mentioned that a loco inspector was on the footplate, perhaps out of concern about the engine's condition.

Also a Derby engine was 'Black Five' 44851, which ran hot on Saturday 5 March on the 6.12am Derby-Bristol express, the loading being a lightweight 186 tons. It came off at Gloucester, causing 29 minutes delay and repairs were effected at Barnwood. No oil was getting to the box because of a defective pipe, something which should have been noticed by the Derby fitter who last checked the engine, so he got reprimanded.

Also in the area that day was 'Royal Scot' 46128 The Lovat Scouts of Crewe North; this class became an everyday sight around Gloucester when dieselisation gave rise to the transfer of engines in the class to sheds like Derby and Saltley.

A familiar class – 3F 0-6-0, a familiar problem – hot box, for 43639 (from 55E Normanton according to the report, actually transferred to Saltley in December 1959, not from 55E, but from 55D Royston), when it got to Cheltenham, the terminating point for its working, the 2.55am class 'K' freight from Washwood Heath, on 9 March. It went to Barnwood for attention. 43639 perhaps was not in good overall condition, being withdrawn in July 1960.

Three unrebuilt 'Patriot' 4-6-0s were transferred to Bristol Barrow Road depot in late 1958 – 45504, 45506 and 45519 – and it was reported at the time that all three visited Stafford Road Works, Wolverhampton, for attention after their arrival on the Western Region. It is probably fair to say they were not the most popular locos with some crews on the Bristol-Birmingham route, although at least one former Barrow Road driver, Tom King, spoke well of them. 45506 *The Royal Pioneer Corps* had the 7.32am Bradford-Bristol class 'A' passenger on 11 March, but failed at Gloucester short of steam, despite running to time. Problems with superheater elements were suspected and the engine went home to Bristol light engine.

'Crab' 42790 of Saltley shed was in trouble with a hot box on the right trailing engine axle on Monday 14 March while working the 1.20pm Burton to Bristol beer train, causing a 79 minute delay for this important traffic. Barnwood could only rustle up a 3F, its own 43427, to take over this class 'C' fitted freight; no doubt it made a good showing – at least there is no further report of hot boxes! 43427, at Barnwood in earlier years, had just transferred back there from the Somerset and Dorset at Templecombe, but stayed only three months before returning to the S&D – perhaps indicating a temporary loco shortage at Gloucester.

4F 44096 had been based on the S&D for years, but on 18 March was heading north light engine, having been transferred to Bury shed, 26D. It got a hot box en route and ended up on Barnwood shed for attention.

Another Saltley 'Crab' with problems was 42758 on 19 March. It suffered bad priming between Kings Heath station, outside Birmingham, and Halesowen Junction while on the 2.45am Washwood Heath-Westerleigh class 'C' fitted freight. The loco came to a stand short of steam and with low water level. So the Barnwood driver called for a fresh engine, which was 8F 48351, the train being held up for an hour.

Some years earlier, on 10 March 1953, 42758 suffered a fused lead plug when the boiler was short of water, while on a similar working, the 5.50am Washwood Heath-Westerleigh class 'C' fitted freight. On that occasion, a Barrow Road fireman received a reprimand for the failure. A 'curious feature of the 'Crab' was the very substantial fall in boiler water level when the regulator was closed ... on steeply graded lines this led to more than average lead plug trouble' according to A J Powell in 'Living with London Midland Locomotives'.

19 March saw another long delay to a freight, the 6.35pm Barnwood to Water Orton. The load was 50 vans, hauled by 3F 43468. When it arrived at Bromsgrove, the left big end was hot due to the brasses being broken and the cotter pin was missing. Sounds like a familiar problem for this class. The delay was 160 minutes with classmate 43389 taking the train forward.

Leeds Holbeck based 'Black 5' 44662 seems to have been a regular on the 7.25pm Bristol-Newcastle class 'A' passenger and mails, actually booked for a Holbeck 'Jubilee', the down working being the *'Devonian'*. 44662 was on report on both 12 and 20 March with injector problems while working the mails. A Millhouses driver also reported similarly on 19 March. On 20 March the Barnwood crew lost 14 minutes to Birmingham, having to take water out of course at Bromsgrove, due to the exhaust injector not shutting off and thus wasting water. Eventually the assistant loco shed master at Holbeck reported: 'Thorough examination of the exhaust injector following several adverse bookings revealed that the water valve was worn. As this cannot be repaired locally a replacement injector has been ordered and the engine retained on local work for the time being.'

On 22 March, 45594 *Bhopal* was in trouble again with Driver F E Ford having another casualty report to fill out for the 8.30am Cardiff-Newcastle express. He wrote, 'When running between Abbots Wood and Spetchley rodding fell from left side of engine, stopped at Spetchley, found that side rod had struck cylinder lubricator rod, breaking off left leading side rod top.' This stop caused 18 minutes to be booked against the loco and a fresh engine took over the train at New Street. Back at Barnwood shed, one of the fitters confirmed he had examined the engine the previous day 'and to the best of my knowledge the lubricator rod and side rod oil cup were secure.'

45506 *The Royal Pioneer Corps* lost 16 minutes between Bristol and Gloucester working the 10.30am Newcastle express on 25 March, being replaced by Northampton shed's 44846 – a rare visitor – at Eastgate. No cause is shown for the delay and the 'Patriot' worked the 1.45pm slow to Birmingham without going to Barnwood shed.

'WD' 2-8-0 90565 came to grief later on 25 March, at Kineton, with the 9.45pm Woodford Halse to Newport

OTHER CASUALTY REPORTS January-March 1960

Date	Loco	Home shed	Train	Problem	Place	Delay (mins)	Notes
19/01/60	42823	21A	1.45pm Gloucester- Birmingham class B	PRI	ER	22	1
20/01/60	44264	85E	WWH - Gloucester goods	VBI	ER		
27/01/60	43949	21A	10.10pm WWH - Westerleigh goods	SOS	GLOS	40	2
27/01/60	48336		11.0am Kineton - Pembrey special freight	SOS			3
30/01/60	43284	21A	7.15pm Gloucester - WWH freight	HOT	ECK		4
01/02/60	44520	21A	4.5am WWH - Gloucester class J	SOS	KH	11	
03/02/60	48220	21A	12.1pm Gloucester - WWH class J	INJ	BNWD	9	5
04/02/60	44209	21A	11.14am WWH - Westerleigh class C	SOS	GLOS	18	
04/02/60	45682	82E	9.15am Paignton - Bradford class A	CWA	GE	7	6
06/02/60	75023	85E	9.23pm Sutton Park - Bristol parcels	SOS	ER		
06/02/60	45649	17A	12.43pm Newcastle - Bristol class A	SOS	GE	3	7
10/02/60	43938	21A	8.55pm Avonmouth - Water Orton	SOS	ER	50	
11/02/60	73000	41B	4.45pm Bradford - Bristol class A	SOS	GE	4	
13/02/60	42761	21A	7.43pm Bristol - Bescot fitted freight	HOT	GLOS	30	8
16/02/60	43963	21A	11.14am WWH - Westerleigh class C	SOS	ER		
17/02/60	73074	41B	10.15am Bradford - Paignton class A	SOS	GE	17	
17/02/60	43925	17A	7.0pm Bristol - Derby class D	VBI	GLOS	10	9
18/02/60	73137		6.52am Gloucester- Birmingham class B	CWA	ER		10
26/02/60	44296	85E	11.14am WWH - Westerleigh class C		WEST		
26/02/60	43975	21A	1.30pm WWH - Gloucester class J	MECH	ER		11
03/03/60	44828	55A	9.5pm Bradford - Bristol class A	SOS	ER		
12/03/60	44662	55A	7.25pm Bristol - Newcastle class A	INJ	ER		12
17/03/60	75004	82E	4.14pm Saltley - New Street ecs	VBI	SALT	12	13
28/03/60	44775	55A	4.48pm Bristol - Leeds class C	MECH	GLOS	23	14

Notes

1	Loco to Saltley mpd
2	Stopped in section Painswick Road - Tuffley; tubes blocked, dirty fire
3	Failed at Briton Ferry (presumably last serviced at Gloucester)
4	Big end cotter missing at Eckington, carried on to Bromsgrove, loco to shed
5	Loco to Barnwood shed, 44263 worked train
6	Train went forward with 73137; 45682 attended to at Barnwood and worked 4.25pm Leicester
7	Fresh loco from Eastgate
8	48507 worked forward
9	44264 worked train; no defect found with 43925
10	Fresh engine from Worcester
11	Left big end cotter missing, new cotter fitted at Barnwood
12	Engine taken off train at Sheffield, to Millhouses shed
13	Repair done, engine back into traffic
14	Actuating steam pipe broken. 44963 took train forward

Key:
COAL: Short of coal; CWA: Carriage warming apparatus ineffective; HOT: Hot box INJ: Injector problem; MECH: Mechanical problem; PRI: Priming; SAND: Sanders not working; SOS: Short of steam; TUBES: Leaking tubes; VBI: Vacuum brake ineffective;

BNS: Birmingham New Street; BNWD: Barnwood; CHEL: Cheltenham; ECK: Eckington; ER: En route; GE: Gloucester Eastgate; GLOS: Gloucester; KH: King's Heath; SALT: Saltley; STOKE: Stoke Works; WEST: Westerleigh; WORCS: Worcester; WWH: Washwood Heath.

Above: 'Patriot' 45519 *Lady Godiva* leaves Cheltenham Lansdown on a down express. This loco, one of three of the class shedded at Bristol Barrow Road, features several times in the casualty reports and apparently had a reputation as a poor steamer.

Alexandra Dock freight. It ran short of water due to the lead plug leaking. It seems also that the gauge glass gave a false reading, as it was broken at the top and a rubber washer was obstructing the waterway. This failure necessitated an extensive repair – one hundred roof nuts in the firebox had to be renewed. The loco had been checked over at Barnwood the previous day by one of the fitters, while the gauge glass had been replaced on 3 March, again at Barnwood, by a different fitter. No blame was attached to the fitters, it was accepted by the Woodford shed master and the Rugby DMPS that the glass was defective. Woodford Halse mpd had been under the impression 90565 was still allocated to Cardiff Canton, so the first report was sent there, only to be informed the loco had transferred to Barnwood in November 1959.

APRIL TO JUNE 1960

'Possibly the cause of this casualty was due to the driver and fireman not being familiar with this class of engine.'

The Barnwood shed master was soon defending another of his fitters after 44171 broke its left valve spindle on 1 April at Cleeve while hauling the 4.0am class 'F' Bath to Landor Street. The Barnwood driver reported: 'When passing between Cheltenham and Cleeve heard very bad knocking. Examined engine at Cleeve and found left valve spindle broken. Proceeded to Ashchurch at reduced speed, engine put inside, changed engines and worked to Landor Street.' The fitter had

checked the engine on 28 March before it worked the 9.45pm Barnwood to Birmingham Exchange Sidings and the shed master said that as the engine had done four days work since then, his man could not be held responsible for the failure. The fresh engine was Ivatt 2-6-0 43047, a regular at Ashchurch on the Evesham and Redditch branch passenger workings.

Later on 1 April, 'B1' 4-6-0 61152 of Millhouses was failed at Stonehouse by a Saltley crew because of 'defective injectors.' It was the assisting engine on the 12.43pm Newcastle-Bristol class 'A' passenger. However upon examination at Barnwood mpd there was nothing wrong with the injectors. The shed master commented: 'This engine is fitted with screw type water feed regulating valves and when it arrived at this depot the valves were screwed right down. Possibly the cause of this casualty was due to the driver and fireman not being familiar with this class of engine.' In which case they were fortunate to have got from Birmingham to Stonehouse before failing the injectors! The following month, on 18 May, 61152 achieved modest fame by being the first reported 'B1' to arrive at Bath Green Park, working the down *'Pines Express'*.

There did not seem any doubt about the state of the injectors on 'Jubilee' 45607 *Fiji*, another Millhouses loco, when it gave up on 4 April at the head of the maximum load – 385 tons – 8.30am Cardiff-Newcastle express. The Barnwood driver reported: 'Whilst working this train the right injector failed at Pirton Sidings (between Defford and Wadborough), worked left injector which was shut off at Spetchley and the clack very bad, unable to draw water to restart it. Stopped at

Dunhampstead with empty boiler, eventually got left injector on, was dragged to New Street.' This episode made for a 65 minute delay booked against the engine.

'WD' 2-8-0s were in the next batch of reports. Barnwood based 90485 apparently failed three times, on 7, 14 and 16 April, but details are sketchy. In any event it did not get to Barnwood until 22 April and appears from the report to have been a recent transfer as its engine history record card arrived at about the same time. The shed master came under pressure from his boss to complete the engine's repair history on the casualty form. The District Motive Power Superintendent even hand-wrote 'Hurry Up!!' on his memo to Barnwood! This loco's allocation here was very brief, being transferred from Cardiff Canton in April and going back there in June, while the other two 'WD's at Barnwood were also transferred away in the summer of 1960, probably not missed by the staff!

Also on 14 April, Cardiff Canton's WD 90069 had a 'mystery' failure at Beckford, on the Evesham-Ashchurch line, running with just a brake van, when the fire was put out with water. It was hauled to Gloucester by 44333, a move seen by the author at Cheltenham. The Gloucester Horton Road driver claimed 'Brick arch needs exam', but scrutiny at Barnwood by the chargehand boilersmith and the running foreman could not find any problems. Indeed they said the brick arch appeared to be in new condition. The shed master did say that the engine had a record of not steaming freely and difficulty had been experienced by Woodford Halse and Gloucester Horton Road men.

Caprotti BR Standard 73144 of 17C, Rowsley, was a casualty on 21 April on the up 'Pines Express'. The right back exhaust cam roller bracket had broken, resulting in the exhaust valve failing to open. The loco was awaiting attention at Barnwood on 24 April. It is interesting to see from the engine history sent by Rowsley shed that 73144 had covered 150,000 miles since emerging brand new from Derby Works on 28 December 1956.

Another 'mystery' failure occurred on 9 May, when a Saltley driver wanted a fresh engine on the 7.40am Bristol-Bradford express. 'Patriot' 45519 Lady Godiva brought the train from Bristol and he claimed it was short of steam. However the Barnwood running foreman could find no reason for this apparent problem. The engine worked the 1.45pm Gloucester-Birmingham and return 5.45pm ex New Street stopper with a Barnwood driver who reported the engine steamed satisfactorily.

A few days earlier, on 3 May, 45519's shedmate 45506 The Royal Pioneer Corps returned home light engine all the way from York. It should have headed the 12.43pm Newcastle to Bristol express, but had problems and the train left York behind 'Crab' 42767.

A type not mentioned to date, Fowler Class 4 2-6-4T 42327 of Saltley got a hot box on 12 May after working the morning

Above: Various 2-6-4T worked Birmingham-Redditch-Evesham-Ashchurch trains. Fowler 2-6-4T 42327, portrayed at Studley, suffered a hot box on 12 May 1960 while on the branch and ended up at at Gloucester Barnwood for repairs. This type was relatively rare at Gloucester.   *P J Shoesmith*

Redditch-Evesham-Ashchurch passenger. It then seems to have gone to Tewkesbury shed for examination, and the return working, the afternoon all stations Ashchurch to Redditch, was taken by Barnwood's 44264. 42327 went to Barnwood shed for repairs and was still there on 18 June, but had gone by 24 June. A further hot box occurred on 12 May with 3F 43359, only noticed when it was on the ashpit at Barnwood after working a class 'J' freight from Washwood Heath. The Gloucester driver candidly admitted: 'Left engine in loco sidings, never noticed defects'. Both leading and driving boxes were reconditioned by Barnwood, where the engine was noted on 22 May. But the old warhorse did not have much working life left as a report from Saltley mpd dated 20 September 1960 notes that: '…this engine has since been cut up.' It had covered 88,167 miles since its last shop repair at Derby in August 1955. On 14 May shedmate 43468 also got a hot box after working a freight from Washwood Heath.

A special working on Saturday 14 May was the 1.10am class 'B' Crewe to Templecombe. Access to the weekly traffic notices would no doubt reveal what this was – a troop train is a possibility as there were furlough specials at this time of year. It did produce a rare loco, 'Jubilee' 45669 *Fisher* whose home depot was 1A, Willesden. Unfortunately the engine failed at Gloucester South Junction with defective injectors, causing a delay of 82 minutes. The engine crew was from Stourbridge, so the train had probably been routed via Woverhampton and Dudley rather than New Street. Barnwood turned out 'Crab' 42799 to take the train forward.

An unusual visitor through Gloucester on 13 and 14 May was Class 'K3' 61853 which double-headed 45662 Kempenfelt from Sheffield on the 12.48pm York-Bristol express on the former date. 61853 returned north on 14 May with a Paignton-Sheffield relief.

Barnwood shed would normally expect a 'Jubilee' off the 8.15am Newcastle-Cardiff, but on 16 May 45639 *Raleigh* failed at New Street. The station pilot, Ivatt class 4 43047, took over, and, with banking assistance from 45639, made a smooth exit towards Five Ways on the heavy 11-coach train.

On 17 May another Willesden engine, not normally an everyday sight, was a casualty at Gloucester. This time it was 'Crab' 42859 in trouble with leaking tubes – six large and 30 small – at the firebox end. Perhaps it had arrived in the area on a working similar to 45669. 42859 had been working the 7.25am class 'E' freight from Westerleigh to Toton when it failed. The train was taken north by Saltley 'Black 5' 44813 while 42859 had its tubes expanded at Barnwood.

More tube problems on 19 May, when 45579 *Punjab* retired at Gloucester off the 6.12am Derby-Bristol with a burst superheater element, being replaced by BR Standard class 5 73160 of 55E Normanton.

Another Derby 'Jubilee', 45610 *Ghana*, suffered a broken right front cylinder cover at Eastgate while on the 1.10am Bristol-Sheffield mails on 26 May. It had only just had an intermediate overhaul at Crewe Works, completed on 12 May, and had since done 2000 miles. The works had not discovered that the cylinder cover was badly flawed around the flange. The loco was noted on Barnwood three days later.

Any bad defects which developed during the first 5000 miles run after shopping had to be specially reported to the Divisional Operations Superintendent.

Something was amiss on 28 May when the down 'Pines' was noted at Yate behind Barnwood's class 4 75002 on 12 coaches – comment was passed by lineside observers that this was more than the maximum load – of 350 tons – allowed for this class! (Class 5s were allowed 385 tons and class 6s 420 tons to Bath and Bristol.)

The 6.35pm Cheltenham High Street goods yard to Glasgow pigeon special empties on Saturday 4 June was booked for two rare locos, as was often the case with these workings. They were Crewe South 'Black 5' 45130 and Carlisle Upperby 'Jubilee' 45703 *Thunderer*, which worked in on the full train the previous day, seen by the author (45703 was not unknown in the area, having passed through Cheltenham almost a year earlier, on 5 July 1959 – while its shedmate, 45596 *Bahamas*, had worked the Glasgow-Cheltenham pigeon special on 5 June 1959). The special was heavily loaded (the equivalent special in June 1962 was booked for 14 vans and 2 van seconds). However the right hand injector of 45130 failed before the return working and the loco remained at Barnwood shed for repairs according to the report (though the author recalls seeing it on Cheltenham Malvern Road shed at some stage in the proceedings). The report notes that the loco had trouble with its exhaust injector a couple of days earlier and received attention at Warrington. 45130's replacement, in the form of BR 9F 92216, meant the train really did have superpower and would have made a fine sight heading north from Cheltenham. Both locos were crewed by Stourbridge enginemen, the special being routed via Dudley and Wolverhampton.

On the two days, June 3 and 4, that the pigeon special was in the area, six 'B1's were also noted – 61004, 61152, 61162, 61181, 61249 and 61312 – showing how common the class had become on various workings. And another Carlisle Upperby engine was seen on 4 June, 45259.

On 5 June 42799 suffered a serious casualty when the front safety plug fused while it was on the 1.33am Sheffield-Bristol parcels. Barnwood examined the firebox and renewed both safety plugs. Class 4 75021 worked the train to Bristol.

Early on 6 June, Barrow Road 'Jubilee' 45662 *Kempenfelt* had charge of the previous night's 9.5pm passenger from Bradford, which lost 35 minutes, mainly due to slipping in the tunnels out of Birmingham New Street station. The Barnwood driver said the sanders were not working and was critical of the train guard who 'had he carried out his proper duties we should not have been there all this time.' The Barnwood shed master amplified this comment when he reported to his boss at Worcester: 'The guard walked from the rear to the front of the train by way of the corridor and, therefore, failed to meet the fireman who walked back in accordance with Rule 178.' Rule 178 states: 'Where it is necessary for the Guard and Fireman to proceed towards each other for the purpose of ascertaining whether the opposite line is obstructed they must proceed along the off side (right-hand side in the running direction) of their train wherever practicable.'

Above: After coming under Western Region control, first as shed code 85E and later 85C, Barnwood received two former GW 0-6-0PT, 7723 and 7756, during 1959. They were usually sub-shedded at Tewkesbury for working the Ashchurch-Upton and Ashchurch-Evesham lines. 7723 was withdrawn in July 1960 and 7788 replaced it. Here 7788, complete with 85C shedplate, and GWR picked out on the tank, passes Hatherley box in 1961 *en route* to Ashchurch.   *R Stanton*

On 8 June GW Collett 0-6-0 2273 of 85A Worcester had the 6.10am Cardiff-Lawley Street class 'D' fitted freight; this was not routed through Gloucester, but via Hereford, Worcester (change engine) and Bromsgrove. Nevertheless 2273 ended up at Barnwood for repair after running hot at Hereford.

4F 44583 of Saltley became a serious casualty, at Hatherley south of Cheltenham on 12 June, while hauling the 9.15am class 'H' Bristol St Philip's to Birmingham Duddeston Sidings, the only regularly timetabled daytime freight on a Sunday on the route. The author saw 44583 standing at Hatherley and noted that the coupling rods had been taken off on one side. A fitter was called to the scene from Barnwood mpd and he found 'the left driving crank pin broken and the axle shifted in the right inside big end web.' There was a two hour delay before the train got underway with a fresh engine, not observed, so it was 40 years later that the author learnt he had missed 16D Annesley's class 'O1' 2-8-0 63610! 44583 was taken back to Barnwood and received new driving wheels and axleboxes, along with reconditioned connecting and coupling rods.

The Summer timetable started on 13 June heralding the season of holiday traffic and lots of extra trains. An excursion on 17 June, 1X68, produced 3D Aston's 'Jubilee' 45696 *Arethusa*, another rare visitor, which failed due to the exhaust injector not working when it returned from a day out at Weston-Super-

Mare. Barrow Road's 73054 took the excursionists back to Wolverhampton from Gloucester.

The addition of summer extras to the regular busy schedules meant that any engine which could turn a wheel was pressed into service. On Saturday 18 June, 4F 44002 of 41D Canklow mpd was on a class 'A' passenger, the 7.30am Newcastle-Paignton, which was basically an extension of the Monday to Friday 10.45am Sheffield-Bristol. 44002 perhaps did well to get to Gloucester before succumbing to a hot box, and retiring to Barnwood for attention, where the author saw it later that day. 75022 replaced it on the train. 44002 was quite likely the rostered engine rather than a replacement; the same shed's 43882 worked the 7.30am Newcastle on 23 July returning north on another relief passenger, while shedmate 44128 worked a Rotherham-Paignton relief that day.

Earlier that week, on 15 June, Barnwood's 7756 – one of its two GW 0-6-0PTs – had been working the Ashchurch to Evesham and return pick-up goods. It appears to have got a hot box, being declared out of service after arrival at Tewkesbury shed. It was subsequently out of traffic at Barnwood from 17 June when the damaged parts were sent to Swindon Works, becoming available for work again on 28 August.

Barnwood's 75002 had the 2.0am Derby-Bristol parcels, P481, on 21 June, but gave up at Eastgate short of steam. 44187, recently repaired at Barnwood after getting a hot box,

OTHER CASUALTY REPORTS April-June 1960

Date	Loco	Home shed	Train	Problem	Place	Delay (mins)	Notes
06/04/60	43853	85E	5.50am WWH-Westerleigh fitted freight	SOS	ER	75	
11/04/60	45263	21A	9.5pm Bradford-Bristol class A	SOS	ER	13	1
14/04/60	43040	21A	1.45pm Gloucester-Birmingham class B	SOS	STOKE	26	2
19/04/60	73166	55A	10.15am Bradford-Paignton class A	SOS	ER		3
23/04/60	90315		8.40pm Woodford-Severn Tunnel Jc goods	MECH	CHEL		
26/04/60	45682	82E	10.15am Bradford-Paignton class A	HOT	GE	7	4
26/04/60	44187	21A	3.45pm WWH-Westerleigh freight	HOT	GLOS	48	5
02/05/60	45597	55A	7.25pm Bristol-Newcastle class A	SOS	BNS	0	6
05/05/60	73054	82E	7.40am Bristol-Bradford class A	SOS	GE		7
09/05/60	42825	17B	4.20pm Burton-Bristol fitted freight	HOT	ER		8
19/05/60	45579	17A	6.12am Derby-Bristol class A	TUBES	GE		9
19/05/60	42823	21A	2.50pm Worcester-Birmingham class B	PRI	21A	9	10
22/05/60	43049	21A	1.35am Sheffield-Bristol parcels	PRI	ER		11
26/05/60	44755	55A	7.25pm Bristol-Newcastle class A	MECH	GE	33	12
28/05/60	45272	21A	10.15am Manchester-Bournemouth class A	SOS	GE		
02/06/60	92150	21A	5.30pm Bristol-Water Orton class D	HOT	ER	8	13
04/06/60	61004		10.30am Bristol-Newcastle class A	HOT	ER		14
05/06/60	45618	17A	9.15am Derby-Bristol class A	SOS	ER		15
11/06/60	45426	9F	10.15am Bradford-Paignton class A	SOS	GE		16
13/06/60	43853	85E	3.15pm WWH-Cheltenham freight	SOS			
14/06/60	43911	21A	5.50am WWH-Westerleigh fitted freight	SOS	GE	25	
21/06/60	73142		9.5pm Bradford-Bristol class A	INJ	BNS	25	17
28/06/60	45607	41C	8.40am Bristol-Bradford class A	INJ	ER	13	18

Notes

1. 44805 worked train from New Street
2. 43040 worked train Worcester - New Street then to Saltley shed, engine changed for return working
3. Loco removed at Gloucester
4. 75023 took train on; 45682 iight engine to Bristol
5. 43855 took train forward; 44187 repaired at Barnwood
6. Fresh engine from Saltley worked north out of New Street
7. 73054 to Barnwood for examination; fresh engine worked train
8. 44520 worked train forward
9. 73160 worked train to Bristol
10. Extra banker needed on Lickey; fresh engine for return working
11. 45682 worked forward from Gloucester
12. Atomiser steam pipe broken; 75009 worked train northwards
13. 92150 to Barnwood for repair; 48607 worked forward
14. Hot box at Dunhampstead, loco worked on to Birmingham at reduced speed
15. 44944 from Gloucester to Bristol
16. Loco last serviced at Leeds; poor coal blamed for failure
17. Station pilot 44775 forward, out of course stop for water at Bromsgrove, losing 25 minutes
18. 75021 took train from Gloucester; 45607 to home depot light engine

took the train on to its destination. As for 75002, ten large tubes were leaking at the firebox end and the boiler inspector took the decision to withdraw it from service. Its estimated mileage was 76,210 since its last general overhaul at Swindon dated 28 November 1957.

Although Barnwood usually dealt with hot boxes, maybe it was too busy to attend to 55E Normanton's 4F 44337 which suffered from its driving axleboxes getting hot on the 7.0pm class 'D' Bristol-Derby on 23 June. The loco was instead sent to Gloucester Horton Road shed for attention.

JULY TO SEPTEMBER 1960

'Right connecting rod badly bent and split. Right cylinder cover broken. Right piston rod bent. Right gudgeon pin broken. Right front coupling rod bent and gradient pin damaged.'

Like 7756, Barnwood's other GW 0-6-0PT, 7723, was usually based at Tewkesbury shed. On 2 July it was employed on the Ashchurch to Upton-on-Severn passenger service for the last time, being condemned later that month. This meant that, with 7756 under repair, the Upton branch workings, Ashchurch shunt and Evesham pick-up turns reverted for a couple of months or so to all-Midland/LMS engine haulage, with 3F and 4F 0-6-0s, also 'Jinty' 0-6-0Ts – including 47465, an 82F Bath Green Park engine, which was working out of Tewkesbury shed from 18 to 29 August, before going to Gloucester and probably home on 30 August. It was not until 19 September that 7756 resumed duty at Tewkesbury, preceded a week earlier by new Barnwood recruit 7788, transferred from 81D Reading.

There was a further glimpse of Midland days when '2P' 4-4-0 40501 worked the 7.42am Gloucester-Bristol stopping passenger on 4 July, possibly covering for a failure or engine shortage. It returned light to resume duty as station pilot at Eastgate. 40501 had only recently been transferred to Barnwood from Bristol Barrow Road, where it had spent time in storage, but its life at Gloucester was brief, being withdrawn the following month. Classmates 40489 and 40540 were in store at Barnwood at this date, having recently been moved from the open into the shed for inspection and a decision on their future. Both saw some further use – the author noted 40489 passing through Cheltenham on 6 August (it was withdrawn from service later in the month) and 40540 was photographed on the Eastgate station pilot duty in September – indeed it was one of the last of the class to be withdrawn, in early 1962.

An extremely rare visitor to the area on 9 July was 12A Carlisle Kingmoor's Pacific 72005 Clan Macgregor which went through Gloucester in the early hours on the 8.45pm Bradford to Bristol – seen by Fireman M Randall who was on the west end pilot, a BR Standard class 4, at Eastgate station. 72005 went back with the 7.45am Paignton-Newcastle. (There was often a west end pilot overnight Friday and through Saturday in the summer timetable to provide assistance up the 1 in 108 Tuffley bank when required for the many heavily-loaded holiday trains heading for the south coast via the Somerset & Dorset or the West Country. This engine faced Bristol, while the regular Eastgate station pilot faced Cheltenham.)

On 12 July, the station pilot at Eastgate, 75021, was called upon to take the 7.40am Bristol-Bradford express, loading to 361 tons, north from Gloucester, following the apparent failure of 'Patriot' 45519 Lady Godiva. 45519 managed to add 40 minutes to the 56 minute schedule from Bristol, with the Saltley driver claiming it was short of steam. Upon examination at Barnwood, the running shift foreman could not find any defect with the engine. The ashpan and fire were found to be clean and the tubes 'very clean'. The foreman commented: 'No obvious cause could be found for this failure but I understand the engine has a bad reputation for not steaming freely.' The DMPS at Saltley added '...no reason to doubt the capabilities of footplate staff.' Lady Godiva was sent back light to Bristol Barrow Road, its home depot. It had covered 14,000 miles since a light intermediate overhaul at Crewe in March 1960. Could it be that a loco's reputation was such that some crews would 'give up' and seek a replacement, with or without justification?

Gloucester men seem not to have had too much of a problem with classmate 45504 Royal Signals on 17 July. It did lose five minutes leaving New Street through the tunnels to Church Road on the 7.5pm Newcastle-Bristol class 'A', but this was put down to the leading sanders not working. The relieving crew at Gloucester, probably Barrow Road men, were happy to take the engine forward to Bristol.

A communications breakdown resulted in a failed engine which was returning light to its home depot being requisitioned for a train working – and promptly failing again. 48419 based at 87F Llanelly was hauling the 5.55pm Washwood Heath to Westerleigh class 'J' freight on 12 July but stopped at Ashchurch with a hot crank pin. 30 minutes was lost examining the engine, which then took the train forward to Gloucester, where a fresh engine came on. At Barnwood it was found that the return crank rod was slightly strained and it was decided to send the engine light back to Llanelly on 13 July with a repair card. But there seems to have been a mix-up between Gloucester Control and Cardiff Control resulting in 48419 being booked by the latter on the 8.50pm Cardiff-Margam freight. It was unable to work that train as Canton depot, where it had gone, found another defect, namely the lubricator connecting rod badly bent. Canton blamed Barnwood for not noticing this fault, but Barnwood denied the loco had the problem when it was there.

A case where the relieving crew did want a fresh engine happened on 23 July. 73000, of 41B Sheffield Grimesthorpe, was working the previous night's 9.5pm Bradford-Bristol class 'A' passenger and had, as usual, a Barnwood crew from Derby to Gloucester. The steam pipe to the steam pressure gauge fractured en route, causing a steam blow in the cab, but the Gloucester man 'did not consider it necessary for the engine to be changed, it was the Bristol driver who relieved him that requested a fresh engine.' 92135 was provided to work the train from Gloucester to Bristol. It was not the only 9F to work a passenger train on the line that busy summer Saturday; Wellingborough's 92127 had train M225.

73000 was noted at Barnwood shed on the Sunday, awaiting attention. Also on that day, a steam blow from the front end developed on Leeds Holbeck's 44983 upon leaving Gloucester station with the 7.25pm Bristol-Newcastle mails, but a stop outside Barnwood depot for a fitter's attention to an auxiliary steam pipe cured the blow and the engine worked forward with just a ten minute delay.

A Saltley crew got the blame when Bath Green Park's 44558 was short of steam on the 9.23pm Sutton Park-Bristol class 'C' parcels on 29 July. The Barnwood foreman commented: 'When engine arrived at this depot, the fire was in fair condition, good quality coal. Mismanagement was probably the cause of failure, more fire under brick arch than at back of fire.' 4Fs performed best with a thin fire under the brick arch and a thicker fire at the back of the box. A Barnwood fireman – who shall remain nameless – discovered in his very first week of firing on the main line that a certain technique was required to get the best from a 4F. He had a Saltley 4F every day on a Gloucester-Washwood Heath freight – and every day, had to stop for a 'blow-up' at Abbot's Wood Junction. After seeking advice from older hands, he became extremely proficient in the art of firing!

Saltley 9F 92070 suffered broken piston rings causing a loss of power while working the 7.0pm class 'D' freight from Bristol to Derby on 3 August; shedmate 45268 worked forward from Bromsgrove after a two hour delay.

Two passenger trains had problems on what was one of the busiest overnights of the summer, Friday/Saturday 5/6 August. 'Crab' 42764 worked the 11.40pm Derby to Paignton, a dated summer extra, but both injectors were defective, and Eastgate pilot 75021 took the train on to Bristol.

On 28 August 1952, 42764 was so short of water that both lead plugs fused. It was in charge of a Barrow Road driver on the 5.50am Washwood Heath-Westerleigh fast goods; the driver received a reprimand for this incident.

The regular 9.5pm Newcastle-Bristol had Derby's 45626 *Seychelles*. Already 30 minutes late leaving Derby, 13 minutes was lost at Burton due to a brake problem on a coach, then the loco was not steaming well, losing 27 minutes to Gloucester.

6 August brought forth a variety of motive power on summer holiday extras, perhaps the most unusual on a passenger working that day was an 8F, 48101, on train M248 Bournemouth-Derby. Also observed were B1 61041 on M244 Bristol-Sheffield; 92057 on the 9.20am Bradford-Paignton; Barnwood's 44264 on 1M21 relief to Nottingham, and shedmate 44296 doubleheading with 9F 92137 on M318 Newquay-Newcastle.

But another of Barnwood's 4Fs suffered a serious casualty this day on a more mundane duty. 43924 had charge of the 5.45pm Birmingham-Bristol stopping passenger. It failed at Bredon when the front safety plug in the firebox fused – commonly known as 'dropping a plug'. This delayed the train for 60 minutes and 45626 *Seychelles*, itself a casualty earlier in the day, came up from Gloucester to assist 43924 and the train into Ashchurch. It left there 72 minutes late according to someone who travelled on it, eventually getting to Gloucester at 9.30pm, 81 minutes late. The 4F had two new safety plugs fitted and 36 crown stay nuts renewed at Barnwood. Disciplinary Form 1 was issued to the Barnwood driver and fireman. This stated: 'When in charge of engine 43924, working the 5/45

Birmingham to Bristol on August 6th, 1960, you failed to ensure that water was maintained in the boiler with the result that one front safety plug was fused and the other damaged, and there was also damage to the firebox of the engine.'
The driver's reply was as follows:
'I do not wish to reserve my defence as after 42 years on the footplate I regret this has happened and still wonder how the damage was done when water was showing in the gauge glass. I also would like to state that we have the least experience booked with us. I quite realise the difficulty of labour shortage there has been, but I still think that the class of work we do demands the most experienced firemen instead of the least. I take full responsibility myself and don't want the blame attached to anyone else. I have always tried to do my best at times under extreme difficulties.'

Opinion amongst footplatemen at Barnwood shed after this incident was that the driver should have stopped as soon as he knew there was a problem maintaining the water level in the boiler, and dropped the fire, rather than carrying on until it was too late to rectify the situation. The matter was settled when the District Traffic Superintendent had a 'suitable coversation' with the men at Barnwood depot on September 14.

Maintaining water in the boiler was a problem for another Barnwood crew with 'Crab' 42823 on 11 August when it worked the 7.0pm Bristol-Derby class 'D' fast goods loaded to 44 wagons. The driver reported 'dirty fire, priming very bad, exhaust injector would not work. Left injector very weak, would not maintain boiler. Arrive Cheltenham High Street 10.27pm, depart 12.2am with engine 75023.'
The Barnwood foreman stated, 'The coal was either from a stack or the bottom of the coal hopper. The firebars were completely covered with hard thin clinker.'

The previous day, 10 August, saw the passage north to Derby Works for scrap of the last Midland Railway Johnson 0-4-4T 'Motor Tank' 58086, hauled by Deeley 3F 43812. 58086 was a Templecombe engine and had worked on the Somerset and Dorset system. Barnwood's last working 'Motor Tank', 58071, was withdrawn in 1956.

On a brighter note, a Barnwood old-timer, 3F 43754, returned ex-works from Derby on 18 August.

Saltley's 43975 lost 18 minutes at Kings Norton on the 7.0am Birmingham-Gloucester on 20 August, not steaming well due to a dirty ashpan, dirty fire and dirty tubeplate. However, it completed the duty and as the Barnwood foreman put it, 'owing to the acute shortage of engine power the tubeplate was cleaned down and the engine put back into service on the 1.18pm Bristol slow.' On that day also, the regular 10.30am Bristol-Newcastle express left Gloucester about 30 minutes late behind Barnwood's 75009, after failure of the booked engine, 45682 *Trafalgar*. 75009 was probably yet again station pilot until called on to head this class 'A' passenger working. 45682 returned light to Bristol for attention.

75023 came to the rescue at Cheltenham on 21 August, this time the problem engine was 21A's 44179 whose tender brakes were sticking.

No less than three 4Fs were in difficulties on 22 August while

working stopping passengers. Derby's 44020 was booked for the 6.52am Gloucester to Birmingham via Worcester, a slow train which called at every station en route except Norton Halt and Fernhill Heath. 44020's small ejector had trouble creating sufficient vacuum after coupling to the train at Eastgate. As the Barnwood driver wrote, 'Had engine ('Black 5' 45253) off Bristol man (at Eastgate) causing a delay of 28 minutes to train. This lost time I had recovered by Barnt Green.' But it was a slow train with some leisurely station stops – 6 minutes at Cheltenham and 11 at Worcester, so it was probably not quite as heroic as it sounded in the report! 44020 had come down on the Saturday with the morning Derby-Gloucester stopper while 45253 also worked in on the Saturday with an evening stopper from Worcester.

Barnwood's own 4F 44167 failed at King's Norton that day on the 5.45pm Birmingham-Bristol slow. The driver reported, 'The train was brought to a stand at Church Road signalbox through the engine being short of steam caused by the large and small tubes leaking very bad in firebox. Control was informed that assistance was required ... to Kings Norton by engine 43911.' 43911 worked solo from there while 44167 went light to Saltley shed, where it was found that six large tubes were leaking. It had received attention at Barnwood on 19 August after similar problems had been reported there.

Meanwhile 21A's 43951 was working home on the 6.30pm Bristol-Birmingham and experienced a problem with the tender brake sticking at Mangotsfield where the driver said, 'I gave it attention with the coal pick' and at Stonehouse. A fitter failed the engine at Eastgate and Ivatt 2-6-0 43047 headed the train north.

'Jubilee' 45626 *Seychelles* which, as mentioned, had problems earlier in the month was still struggling on 24 August. It had the 12.43pm Newcastle-Bristol express, actually running pretty much on time, but gave up at Gloucester with dirty smokebox, ashpan and fire, as well as a row of bricks missing from the

brickarch. Classmate 45598 *Basutoland* took over, but also seems to have had steaming problems on the journey to Bristol and again the next day on the 12.40pm Bristol-Bradford, the up 'Devonian'. 45598 had originally been booked for the 4.25pm Gloucester-Birmingham stopper on 24 August, but while being serviced at Barnwood was found to have a broken tender spring.

30 August sees the first mention of the Bromford Bridge oil trains, which ran from Avonmouth and from Fawley refinery near Eastleigh. These were recent additions to traffic on the line, not being shown in the working timetable for the period, although they featured in the one which commenced on 12 September 1960. These were heavily loaded trains and needed big engines, hence were rostered for BR's most modern steam locomotives, 9F 2-10-0s. Bristol Barrow Road had some of the class allocated there in August especially for these workings. One was 92248 which had the 11.40pm Avonmouth-Bromford Bridge, with 29 tanks, equivalent to 83 loads, but struggled, losing no less than 191 minutes on the way to Gloucester – the schedule from Avonmouth to Gloucester South was 96 minutes, so it had trebled the running time! It arrived on Barnwood shed at 5.30am having given up. The foreman's inspection found, 'Fire dirty, irony clinker also fire grate required cleaning off, no air space.'

No doubt Barnwood did what they could before the engine went off shed again at 9.30am to take over the similarly loaded 12.25am Fawley-Bromford Bridge. It lost 20 minutes between Cheltenham and Spetchley, due again to shortage of steam. After this turn, it went back to Barrow Road for a water test and tube examination. All this with a loco which was brand new just eighteen months ago.

30 August also saw the return north of ex-LNW 0-8-0 49452 of 21C Bushbury which had worked a Washwood Heath-Westerleigh freight the previous day.

9 September 1960 saw the last visit to Tewkesbury shed of a

OTHER CASUALTY REPORTS July–September 1960

Date	Loco	Home shed	Train	Problem	Place	Delay (mins)	Notes
01/07/60	45610	17A	11.45pm Bristol-Derby parcels	SAND		20	1
02/07/60	43986	21A	9.15pm Bath-Lawley Street class D	VBI	ER	15	
06/07/60	42847	17B	4.20pm Burton-Bristol class C	HOT	ER	3	2
09/07/60	44226	21A	9.15pm Bath-Lawley Street class D	SOS	ER	30	
11/07/60	45618	17A	6.12am Derby-Bristol class A	PRI	ER	17	3
17/07/60	73065	41C	7.25pm Bristol-Newcastle class A	SOS	ER	10	4
19/07/60	48351	21A	3.45am Westerleigh-Water Orton class D	MECH	GLOS	76	5
19/07/60	45076	26A	2.0am Derby-Bristol parcels	INJ		52	6
23/07/60	92155	21A	4.30pm Bristol-Water Orton class D	SOS	ER	11	
01/08/60	45597	55A	7.25pm Bristol-Newcastle class A	HOT	CHEL	63	7
10/08/60	45618	17A	Cardiff-Newcastle class A	PRI	ER	13	8
10/08/60	43853	85E	7.25am Westerleigh-Toton class E	MECH	GLOS	15	9
22/08/60	75009	85E	6.40pm Park Lane-Birmingham class C	SOS	DERBY	20	10
25/08/60	44490	21E	2.50pm Worcester-Birmingham class B	SOS	ER	20	11
26/08/60	73068	82E	9.45am Bournemouth-Manchester class A	VBI		70	12
28/08/60	44667	15C	9.5pm Newcastle-Bristol class A	SOS		25	13
01/09/60	43949	21A	8.55pm Avonmouth-Birmingham freight	SOS	GLOS		
09/09/60	75009	85E	7.5pm Sheffield-Bristol class A	MECH			14
11/09/60	73043	41B	9.5pm Newcastle-Bristol class A	VBI	ER	22	15
20/09/60	43911	21A	5.30pm Bristol-Water Orton class D	SOS	ER		16
23/09/60	44334	17A	2.0am Derby-Bristol parcels	SOS	ER		
24/09/60	43924	85E	6.30pm WWH-Westerleigh freight	MECH			
26/09/60	90685	86A	2.55am Cardiff-Woodford freight	SOS	ER		17
27/09/60	73000	41B	8.40am Bristol-Bradford class A	SOS	GE	13	18

Notes

1	Engine slipped to a stand at Kingswood Junction, sanders not working. Assisting engine called for.
2	Fresh engine from Gloucester
3	17 minutes lost from Gloucester to Bristol due to priming
4	Loco changed at New Street, unable to maintain boiler pressure
5	Driver's brake valve blowing through, unable to release engine brakes. 61041 took over train
6	Failed at Coaley Junction. 44818 took over
7	45618 sent by Barnwood to take over at Cheltenham
8	Brick arch collapsed; train also overloaded for engine; fresh loco from New Street
9	Left front big end bolt missing. 44264 took over train
10	75009 had a burst tube; 45280 worked train
11	44490 to Saltley for inspection. 43855 took return 5.45 class B to Gloucester
12	73068 failed at Selly Oak, light to Saltley. Unknown 4F took over
13	44667 failed at Duddeston. 73065 off Saltley shed took over
14	Loco failed at Westerleigh
15	73051 took over at Gloucester
16	Loco to Barnwood for inspection
17	Loco stopped at Malvern Road shed to clean fire
18	Fresh engine from Gloucester. 73000 to Barnwood for servicing; later worked 4.0pm Bristol class B

loco off the 7.55am Washwood Heath-Redditch-Evesham freight, which ran light from Evesham for servicing. This day, it was Saltley's Class 5 45088, but the previous day it had been 0-8-0 49106 of 21B Bescot. On this turn the loco worked home on the 5pm Ashchurch-Lawley Street freight, again via Evesham. From 12 September both workings were cut back to Redditch.

Standard '5MT' 73136 suffered a serious failure at Haresfield on 21 September while working the 7.0pm class 'D' Bristol to Derby, 5M27. The right hand connecting rod became detached from the crosshead. The shed master's report lists the following damage: 'Right connecting rod badly bent and split. Right cylinder cover broken. Right piston rod bent. Right gudgeon pin broken. Right front coupling rod bent and gradient pin damaged.'

Presumably this caused all manner of mayhem and delay on the line, as it was six hours before the train got under way again, with 4F 43911. 73136 stood at Barnwood, while the damaged parts were renewed at Derby. It was 27 October before the loco was put back into service.

OCTOBER TO DECEMBER 1960

'I wish to report having to change engines again at Bromsgrove this week...'

Barnwood's own 44209 was specially proposed for a visit to works after its right cylinder cover broke. This occurred between Lifford and Hazelwell when it was hauling the 9.45pm Barnwood to Birmingham Exchange Sidings class 'D' freight on 7 October. The Barnwood driver stated, 'the engine developed a bad knock on the right side after passing Lifford, this got

worse, so stopped ... and found that the right cylinder cover had gone. Delay to main line was from 11.55pm to 12.40am.' The casualty report noted the cause was 'probably due to piston head breaking and broken parts damaging cylinder casting and covers at both ends.' The Barnwood shed master did not fault anyone for the problem. 44209 had covered 13,100 miles since a heavy intermediate overhaul at Derby in May 1960.

43013 of Saltley had the 11.38am Redditch-Evesham-Ashchurch passenger on Monday 10 October, crewed, as rostered, by Tewkesbury men. On arrival the engine went to Tewkesbury shed for servicing before working the 4.39pm to Redditch (this applied Monday to Friday; on Saturdays there was an additional round trip Ashchurch-Redditch-Ashchurch to fill the layover period). At Tewkesbury 43013 was found to have defective firebars, so was failed. This working was Saltley engine turn 27 and officially diagrammed for a class 4 tank, but the Ivatt class 4 2-6-0s were regular performers as well. There is no note of what took over from 43013, it could conceivably have been 7788 or 43754, which were on other Tewkesbury shed turns, or perhaps a fresh loco came up from Gloucester.

When the famous Ashchurch flat crossing was extant, the whole Redditch train – engine and coaches – used to turn on the triangle formed by the Evesham branch, the Tewkesbury branch and the direct link between the two across the Birmingham-Bristol main line. The loco was usually a tender engine, such as a '2P' or Compound. After the crossing was removed in 1957, turning was no longer possible, so it became more desirable to have a tank loco on the working, or a tender engine such as the Ivatts, which were relatively acceptable running tender-first for some distance.

The 11.38am Redditch-Ashchurch passenger working had smaller than normal power on Saturday 15 October with 85F

Above: Ivatt 2-6-0s including 43013, seen at Alvechurch, were also regular performers on the Birmingham-Redditch-Evesham-Ashchurch route. 43013 was unable to complete its turn on 10 October 1960 due to defective firebars, discovered when it went to Tewkeesbury shed after working along the branch to Ashchurch.

P J Shoesmith

Bromsgrove's 'Jinty' 3F 47308 deputising for the class 4 tank.

43013 was back in action by 17 October when it was on the 5.10pm New Street-Ashchurch passenger, also working the turn for the following two days.

7788 was on the Ashchurch-Upton passenger on 15, 17 and 18 October, but then returned to Barnwood for ejector box repairs. It was back on branch duty on 7 November, its turns having been covered in the meantime by either 7756, 43645 or 43754.

'Patriot' 45506 *The Royal Pioneer Corps* appears to have been regular power for M321, the 7.5pm Newcastle-Bristol mail train in mid October. It worked from Sheffield, but on both 14 and 17 October, failed at Gloucester short of steam. 45506's condition did not improve; one Saturday before Christmas 1960 there is a record of it having a very poor run out of New Street on the 10.28am Sheffield-Bristol.

Sunday trains were often heavily delayed due to engineering works, and with late running causing missed connections, a 'Q' – as required – passenger train, the 6.42pm Birmingham-Bristol was put on, usually just four or five coaches with the New Street station pilot. On 23 October this ran with Saltley 2-6-0 43040 and a 180 ton load. But with very dirty tubes and tubeplate, 43040 expired at Gloucester – more delay for the long-suffering passengers!

Classmate 43047 was short of steam next day on the 1.45pm Gloucester-Birmingham stopper, it lost 30 minutes to Bromsgrove where it was replaced by 4F 44091. Normally the engine off this train worked back on the 5.45pm Gloucester stopper from New Street, but 44091 had problems warming the carriages, so the Barnwood crew transferred to recently ex-works 73090, of 84G Shrewsbury, off Saltley for their return – three engines instead of one!

The driver, F E Ford, and his fireman, T Phelps, had a rough week on this turn, as on 27 October the driver wrote, 'I wish to report having to change engines again at Bromsgrove this week caused by engine 42824, 17B, priming very bad. Time lost Worcester to Bromsgrove 10 minutes. Time lost changing engines at Bromsgrove 20 minutes. Engine 42824 changed for 43040. Changed 43040 for 44465, 17A, at Saltley. Lost 10 minutes with 44465 New St. to Blackwell (on the 5.45pm), very dirty tubeplate.' Again three engines instead of one!

42824 was one of five 'Crabs' fitted with Reidinger rotary poppet valve gear and spent time at Rugby Locomotive Testing Station where it did 47 runs during performance testing on the valves and another 68 when the valves were modified. Incidentally all five were 17B Burton engines in 1960, having been Saltley allocated when they were fitted with the aforementioned valve gear.

BR Standard '5MT's suffered minor avoidable delays on 29 October. 73054 of Bath mpd was on the 7.40am Bristol-Bradford, but was replaced at Gloucester by Millhouses' 73046. The cab of 73054 filled with steam due to a slacked back nut on a pipe fitting, easily rectified according to the correspondence, but not easily noticed by the Barrow Road fitter who checked the engine before it left shed according to his shed master.

A little later that day, another Bath engine, 73028, was working home on the down 'Pines Express' but was failed at Gloucester allegedly short of coal, having 1 ton left in the tender to complete the journey. Gloucester Control was not impressed, putting the 7 minute loss changing engines down to 'slow moving of the drivers.'

One engine which slipped through that day without incident was 'K3' 61808 on a Washwood Heath to Westerleigh freight.

31 October found another '5MT' having problems, this time Derby shed's Caprotti valve gear 73135 at Cheltenham, when the universal joint from the gear box to the valve drive became disconnected. A delay of 31 minutes ensued to M322, the 5.0pm Bristol-York express. It was a touch ironic that the replacement engine was another Caprotti, 'Black 5' 44756. 73135 had achieved some fame on 15 October by working over the S&D from Bath to Bournemouth, following another loco's failure at Bath; it was believed to be a possibly unique appearance of a Caprotti-fitted member of the class on the line.

Shortage of coal caused 43122 to give up its working on Sunday 6 November at Gloucester, the 1.33am Sheffield to Bristol parcels. The Barnwood foreman remarked 'In view of the heavy loading of this train the amount of coal on the tender – 10 cwts – would have been insufficient to work through to Bristol.' Although it was a Saltley engine, 43122 had worked this train through from Sheffield, so possibly a relief loco had been anticipated at Birmingham.

A Barnwood driver got the blame for the failure of 'Crab' 42846 on 11 November. He claimed the right hand injector was defective. A fitter's examination found the trouble was caused by an accumulation of coal around the tender water feed handle preventing it from being fully open. The DMPS wrote, 'In my opinion this defect could have been rectified by the driver through a little more investigation and this delay and change of engine could have been avoided. I am asking my Barnwood shed master to follow up accordingly with the driver.' The delay was 44 minutes on the working, the 11.20am class 'E' Bristol-Washwood Heath, which 48669 took forward from Gloucester Engine Shed Junction.

On 13 November Barrow Road 9F 92231 had charge of the 11.40pm Avonmouth-Bromford Bridge with 25 loaded petrol tanks. At 4.12am on the 14th, the engine parted from the train. Barnwood driver L Grant wrote, 'Running between Brickyard Crossing and Landor St (Birmingham) with above train, engine became parted from train, coupling pin came out of tender screw coupling which had become strained, also vacuum hose pipe was damaged. This no doubt was caused by the rapid action of brake on train which I found very difficult to regulate when descending steep gradients. I arranged with Control to send another engine to assist me, for braking purposes, to work train forward to Bromford as a non-fitted train. Note the Controller at Birmingham informed me that this same engine had parted from trains three times recently.' Just as well that the brakes worked – a runaway train of loaded petrol tanks in Birmingham hardly bears thinking about! The ruling gradient in the area of the incident is 1 in 62.

'Black 5' 44945 had sole charge of the 8.30am Cardiff-Newcastle express on 14 November, with a train load of 377 tons, against a regulation maximum load of 350 tons. The driver

Above: The Fawley to Bromford Bridge 'Esso' tanker train service was important traffic for BR, but performance of the booked '9F's often gave management cause for concern. The problems seem to arise mainly from a lack of adequate servicing and a shortage of coal capacity for engines working throughout. Some '9F's were allocated to 71A Eastleigh depot on the Southern Region especially for these trains, and here is 92231 with a 71A shedplate clearly visible, heading the empty tanks back to Fawley during 1961. The loco has lamp brackets fitted halfway down the smokebox, – necessary for carrying Southern Region disc headcodes. Note the brake van and 'runner' wagons which were marshalled at either end of the train, creating a gap between the locomotive and the highly inflammable cargo.

R Stanton

blamed a delay on the overload and the engine not maintaining a full head of steam due to inferior coal. The Barnwood shed master refuted the remark about inferior coal, saying it had been supplied with three tons of Arley grade 2A, the highest grade allocated to the depot 'and no previous complaints have ever been received regarding its steaming qualities.'

3Fs have not been mentioned recently, but 43521 came to grief on 24 November at Spetchley with the right hand big end hot. The metal had fused and, upon going for repair, the brasses were remetalled and refitted. The failure was due to lack of lubrication, blamed on the preparation driver at Barnwood, who claimed the engine had been properly serviced. Its train, the 7.15pm Gloucester-Washwood Heath class 'H' freight was taken on by Bromsgrove shed's sole 3F 43762.

48669 which had itself taken over from a failed loco on 11 November was in trouble on 1 December. Although being a Saltley engine working from Toton to Birmingham with a Burton crew, a copy of the report landed on the Barnwood shed master's desk because the defective parts, namely the exhaust injector and live steam injector, had last been examined at Barnwood on 30 November. The failure caused a 139 minute delay at Branston and another 8F, 48606, took over the working. The Gloucester fitter got the blame for not checking the injectors properly.

92231 was in the wars again in the early hours of 2 December on the 11.40pm Avonmouth-Bromford Bridge. It lost 115 minutes in running between Ashchurch and Barnt Green, owing to '...inferior coal, dirty fire and tubes blocked up.' The scheduled time between those two places was 75 minutes, including a 10 minute stop at Bromsgrove for water and attachment of bankers, also 10 minutes recovery time, so it really was 'doing rough'.

It was a 3F to the rescue on 7 December at Stonehouse on the 7.42am Gloucester-Bristol stopper. 43284 took over from 44226 whose ashpan was warped preventing the damper opening. The delay was 45 minutes. This time the preparation fireman got a 'rocket' for not testing the dampers prior to the engine leaving shed. If, as seems likely, 43284 was not steam-heat fitted, the passengers would have had a cold journey.

A notable arrival at Gloucester later in the day was a 'Battle of Britain' Pacific, 34053 Sir Keith Park, on the up 'Pines Express'. This highly unusual occurrence was due to a landslip on the S & D, causing a diversion of the service via Salisbury and Bristol for several days. But 34053 went no further, being replaced by the engine of the 1.45pm Eastgate to Worcester slow. The Bath Running Shed Foreman's log indicates that 73047 was specially sent to Bristol with Bath men to go north on the up 'Pines' that day. He also stated: 'West Country engine not allowed over the road' (Bristol-

Gloucester). So much for the best-laid plans! But two days earlier 'West Country' 34102 Lapford managed to get through to Birmingham New Street on the same working, returning next day on the down 'Pines'. The train also had unaccustomed power on 6 December in the form of 4942 Maindy Hall which also went through to New Street and returned next day. (Peter Smith has written more about these workings in his book 'The Somerset & Dorset From The Footplate').

However there were other occasions when Southern 'Pacifics' worked to Gloucester. Mike Randall recalls seeing an unrebuilt one on Barnwood depot which had worked up from Bath and then returned on the 12.20am Gloucester Eastgate-Westerleigh class 'H' freight, rostered for Barnwood men. Apparently this was not an isolated instance.

On 8 December, 45607 *Fiji*, hauling the 5.15pm Sheffield-Birmingham class 'A' passenger, had similar problems to 48669, namely a defective exhaust injector and an unreliable live steam injector. Again, the last inspection of these fittings had been at Gloucester, two days previously. The fitter – 'a very reliable man' according to Mr Cole the Barnwood shed master – denied having damaged the joint face of the injector combination cone. This was accepted by the Millhouses loco shed master, where 45607 was based. The BR paper trail meant a copy of the report went to the line traffic manager at King's Cross. It is difficult to believe he would have much interest in a train going from Sheffield to Birmingham!

43040 was no stranger to failure and got short of steam again on 12 December; the driver's report is copied on page 31.

4F 43924 had major problems again on 16 December, after the drama of dropping a plug in August. It was in charge of the 1.45am class 'H' Westerleigh-Washwood Heath, with a Barnwood crew, when it suffered a serious mechanical failure at Tuffley with a lack of lubrication due to the left-hand big end syphon top being missing. The exact details of the damage, all on the left-hand side, were, 'Big end brasses broken up and missing. Connecting rod damaged at large end. Gudgeon pin bent. Slide blocks broken. Piston rod bent. Big end journal damaged.' All the parts mentioned were sent to Derby for repair or replacement. The Barnwood shed master wrote, 'As this loco had only worked from Bristol the left-hand big end syphon top must have been missing or slack before the engine left shed and, I consider, should have been detected by the preparing driver. The right-hand big end was also found to be completely dry and the metal had started to fuse.'

The following day, the rather forlorn 43924 shared the shed with an unusual visitor to Barnwood, class O4/3 2-8-0 63754.

Pulling a heavy train through the tunnels out of Birmingham New Street from a standing start needs working sanders, especially in the middle of December. 'Jubilee' 45663 *Jervis* lost 23 minutes slipping in the tunnels on the 366-ton 12.15pm York-Bristol express on 18 December. The Barnwood driver mentioned that the engine had already lost 20 minutes through the same problem before it got to Birmingham, also the loco had been reported previously for the same defect.

Barnwood's 44272 had a light intermediate overhaul at

Derby Works dated 9 November and had since covered only 354 miles when it failed at Chaddesden on 20 December with the 2.50am class 'J' to Water Orton. The boiler pressure gauge burst internally, but a smart piece of work saw a replacement fitted 'whilst standing in traffic at London Road Junction.' A copy of the casualty report went to the running and maintenance officer at Paddington – wonder what he made of it!

The next casualty occurred due to an engine not keeping to its diagrammed turn. The booked locomotive for the 8.30am Cardiff-Newcastle express (10.5am ex-Gloucester) on 21 December, 45274 of 55A Leeds Holbeck, mysteriously disappeared. One can only speculate as to what happened to 45274 after it worked the 4.5pm Bristol stopper from Gloucester on 20 December, per the loco's diagram – there is no report of it failing, so the presumption must be that it was 'borrowed' by Barrow Road for another turn, perhaps a relief passenger or parcels train bearing in mind this was just before Christmas. Barrow Road's own Standard class 4MT 75004 brought up 45274's booked return, the 9.48pm stopper ex Bristol.

45274's next diagrammed turn was the aforementioned Newcastle express, for which 75004 was officially underpowered, being only a class 4 loco. Derby depot's 45648 *Wemyss* was the substitute provided by Barnwood – it should have worked the 2.34am Sheffield parcels from Gloucester, but was held back for the Newcastle. The 'Jubilee' had problems with shortage of steam, losing 29 minutes and being replaced at New Street. In his casualty report, Mr Cole the Barnwood shed master noted: 'During examination of 45648 prior to leaving the shed the smoke box was found to contain a large amount of char and the bottom row of tubes completely blocked. The brick arch was also found to be defective with half the brick missing and a hole in the remainder. 'The Gloucester driver who brought 45648 to shed the previous evening at 11.15pm off the 5.15pm ex-Sheffield got a memo from Mr Cole asking him to explain why the poor condition of the engine had not been drawn to the foreman's attention. The driver replied: 'The amount of char left in the smoke box would not have stopped the engine from steaming to such an extent as to warrant changing at New Street. Neither would the condition of the brick arch. The main cause (for the failure) was the engine being prepared for the 2.34am parcels and being coaled with second grade coal approximately eight hours before (the 10.5am).'

There were no doubt other knock-on effects of these changes – Millhouses would have been expecting 45648, a class 6P 4-6-0, off the 2.34am Gloucester-Sheffield parcels on 21 December, but instead got Ivatt class 4MT 2-6-0 43046, which took the train from Gloucester, unless this was replaced en route. (The Bath shed foreman had a similar problem on 23 December when a BR Standard class 4 arrived there on the 12.37am parcels from Leicester, booked for a bigger engine, as the return working was the up 'Pines').

The concluding report is copied here. The observation that 'It is the practice ... to avoid booking this (8.6am ex-Sheffield) engine on the 4.5pm Bristol so that the fitters and boilersmith can thoroughly examine it ...' may have come about after the DMPS' comment back in January – 'I am asking the Gloucester Barnwood shed master to have special attention given to the servicing and coaling of the engine working this important (8.30am Cardiff-Newcastle) service, with special reference to the tubes and tubeplate.' It is somewhat ironic that if Barnwood had followed its usual practice and kept back 45274, the Newcastle train would probably have had a trouble-free journey.

It is interesting to note that neither of the Millhouses turns mentioned in the concluding report were being worked by Millhouses engines. 45274 and 45648 were both at Barnwood shed on Sunday 17 December so were probably the diagrammed engines for the Sheffield parcels and Newcastle express during that week.

Barnwood men had a tragic history with 45274. It was the loco hauling the 12.15pm York-Bristol express on Sunday 23 January 1955 which derailed at Sutton Coldfield due to excessive speed. 17 people were killed, including the Barnwood fireman and Burton driver on the engine. The train had been diverted due to track maintenance on its usual route at Tamworth, and while the pilotman for the diversion drove the loco, the Barnwood driver was travelling in a carriage, strictly against regulations. It is understood that the Barnwood man never again worked main line trains. 45274 was shedded at Bristol Barrow Road at the time.

45648 was not the only 'Jubilee' in trouble on 21 December. Barrow Road's 45682 *Trafalgar* lost 25 minutes between Bristol and Gloucester on the 2.15pm express to York, with defective elements and exhaust injector not working. 'Black 5' 45269 took the train forward from Eastgate, while 45682 went back light to its home depot for a water test.

Just over a week later, on 29 December, 45682 again only got as far as Gloucester where it gave up short of steam on the 7.40am Bristol-Bradford express. The elements were still blowing and it went back light to Barrow Road for attention, while 44962 took the working north.

Also on that day, another 'Jubilee' 45594 *Bhopal* was struggling on the 5.50am Washwood Heath-Westerleigh fast freight. It was short of steam, causing a delay of 57 minutes, and 44841 replaced it at Gloucester. 45594 went back to its home depot, 41C Millhouses, light engine.

An unusual loco deputising on the 5.45pm Gloucester-Bristol local on 29 December was Horton Road's Collett 0-6-0 2245. This train should have had the engine off the Newcastle-Cardiff, due in Gloucester at 2.34pm, booked for a Derby 'Jubilee'. It would be interesting to know if 2245 took the next part of the working, the heavily loaded 11.45pm Bristol-Derby parcels.

The last report for 1960 was in the early hours of 31 December and once more featured the 11.40pm Esso tanks from Avonmouth to Bromford Bridge. It was not so much the engine, 9F 92004, which caused the 60 minute delay, at least in management's eyes, as the Barnwood driver who brought the engine off the train to the shed to clean the fire and take coal. The shed master was not impressed by the driver's actions and wrote, 'As this engine had only worked from Bristol, I do not consider this was necessary.'

Above: In August 1958 Barnwood received a couple of BR Standard class 4 4-6-0s, 75009 and 75023, the most modern locos to be shedded there at that time and intended to replace 4-4-0s. 75002 also came to Barnwood later. As well as assisting express passenger train workings like the one shown here, with 75023 and 44691 heading north, the Standards worked station pilot duties at Gloucester Eastgate when they would often be called upon to take over from an ailing loco on an express train. There are numerous examples in the casualty reports. In turn, the '75's were replaced by BR class 5 Standards 73091, 73092 and 73093.
R Stanton

BRITISH RAILWAYS

M.P. .... Depot .... Region. Date .... 21.12.1960.

Date of Incident ...... 21.12. 19 60   Place of Incident .... Glos. - New St.

Driver } * W. Mellor   Reg. No. 16   Depot   Glos.
Motorman   209   do

Fireman } * Ford   Reg. No. ....   Depot ....
Second Man

Guard ....   Station ....

10.5   a.m. Passr. } * Train from ...... Glos.Cen. .... to .... New St.
    p.m. Frt.

11 Veh.   Actual load   277

on ...... 21.12. ...... 19 60   Reg. load ......   Train   Assisting

Locomotive/Diesel/Power Car } * No. 45648
Electric Multiple/Diesel Elec./Unit.   6P

Class/Type* ....

*Delete words not applicable

Time lost by Loco 29 mins.

Time was lost by Loco to the extent of 29 mins owing to being
short of steam, Inspector Smith of Worcester will supply particulars,
we ~~xxx~~ wired for fresh eng. at Dunhampstead.

                           W. Mellor.

              Signature

Engine 45648 (17A) Non-Mechanical Casualty –
En route 21.12.60. 8.30 am. Pass. – Cardiff to Newcastle.

CONCLUSION.

The engine diagrammed to work this train is Millhouses Turn
Nos. 6 and 7 which arrives on my Barnwood Depot at 2/0 pm. off
the 8.6 am. ex Sheffield the previous day and is booked to work
the 4/5 pm. Pass. to Bristol returning to Gloucester with the
9/48 pm. ex Bristol to be stabled until it works the 8.30 am. ex
Cardiff forward from Gloucester.

It is the practice, whenever possible, to avoid booking this
engine to work the 4/5 pm. Bristol so that the Fitters and
Boilersmith can thoroughly examine it and carry out any repairs
or servicing duties that may be required.

On 20.12.60, however, there was no other engine available
at Barnwood for the 4/5 pm. and it was necessary to use the
diagrammed engine 45274 (55A). It was examined by the Barnwood
Chargeman Boilersmith and an entry was made on the Tube
Cleaners sheet for the large and small tubes to be cleaned when
the engine returned to shed off the 9/48 pm. ex Bristol, in
readiness for the working of the 8.30 am. Newcastle the following
morning.

Engine 45274, however, did not return with the 9/48 pm.
ex Bristol and when the engine off this train arrived on
Barnwood shed at 1.10 am. it was found to be in B.R.Class 4 M.T.
No. 75004. The only suitable engine then on the Barnwood shed
to work the 8.30 am. ex Cardiff was 45648 (17A) which had worked
in off the 5/15 pm. Pass ex Sheffield and had been prepared for
the 2.34 am. Sheffield Parcels which was its booked return
working.

As nothing had been booked against the steaming of this engine
by the Driver off the 5/15 pm. Sheffield it was put to work the
8.30 am. ex Cardiff without further attention being given.

                            For J. POWELL

OTHER CASUALTY REPORTS October-December 1960

Date	Loco	Home shed	Train	Problem	Place	Delay (mins)	Notes
05/10/60	44226	21A	5.0am WWH-Westerleigh freight 9V21	MECH	GLOS	56	1
05/10/60	44296	85E	11.25am WWH-Bristol freight	HOT	GLOS	27	2
06/10/60	45659	55A	9.2pm Newcastle-Bristol class A	SAND	BNS	5	
13/10/60	44203	21A	5.0am WWH-Westerleigh class J	HOT	GLOS	25	3
07/11/60	44143	21A	8.50am Bristol-WWH goods	SOS	GLOS		
03/11/60	43013	21A	1.45pm Gloucester-Birmingham class B	SOS	ER	9	4
13/11/60	45656	41C	12.15pm York-Bristol class A	PRI	ER	25	
19/11/60	44179	21A	7.43pm Bristol-Bescot class E	SOS	ER		5
24/11/60	45682	82E	12.48pm York-Bristol class A	SOS	ER		6
08/12/60	42823	21A	2.45am WWH-Westerleigh fitted freight	SOS	ER	24	7
09/12/60	43041	21A	4.6pm Birmingham-Bristol special passenger	SOS	GE	5	8
12/12/60	43040	21A	2.50pm Worcester-Birmingham class B	SOS	ER	13	9
15/12/60	44263	21A	2.45am WWH-Westerleigh fitted freight	SOS	ER	31	10
15/12/60	42855	17B	2.50pm Worcester-Birmingham class B	CWA	WORCS	50	
16/12/60	45598	17A	12.43pm Newcastle-Bristol class A	MECH	GE		11
18/12/60	45562	55A	9.2pm Bradford-Bristol class A	SAND	BNS	5	

Notes

1. Piston gland blowing, loco returned light to Saltley. 42758 took train forward
2. 43887 took train on. 44296 repaired at Barnwood
3. 44045 took train. 44203 repaired at Barnwood
4. Loco to Saltley mpd for attention to blocked tubes
5. Failed at Cheltenham High Street, examined at Barnwood, tubes badly blocked
6. Loco taken off at Gloucester.
7. Dirty tubeplate and heavily clinkered fire. Stopped for blow up at Hazelwell. Fresh loco from Gloucester
8. 43041 returned light to Saltley. 75002 took train to Bristol
9. Ashpan full when checked at Barnt Green. Fresh loco for 5.47pm return class B
10. 42846 worked from Gloucester. 44263's tubes cleaned and returned to Saltley
11. 42904 worked train to Bristol

DRIVER'S/MOTORMAN'S REPORT...

Time lost by Loco - short of steam.

I have to report that I took to 43040 at Worcester at 3/15pm and left at 3.19pm or 29 mins late, we had signs ~~the the 3.15pm and left at 3.19pm~~ that we would be short of steam quite early and when we arrived at Blackwell we had just over 100 lbs and the vacuum was failing. We had about 1 in of water in the glass. I proceeded to Barnt Green and stopped under the protection of signals and went to the signalbox and asked for a fresh eng. from Saltley for the return journey (5.47 ex New St) My Fireman opened the hopper in the presence of the Station Master Barnt Green who would say it was full up, in passing I should mention that the gauge lamp was also empty, we arrived New St at 4.35 or 45 mins late. This makes time lost by loco about 13 mins.

Signature .... W.Mellor.

CHAPTER FOUR

REPORTS FOR 1961

Barnwood's shed code changed from 85E to 85C on 1 January, but things were set to change in rather more dramatic fashion during the course of the year. With increasing motive power modernisation on other routes, it was inevitable that diesels would start operating on the Birmingham-Bristol main line. It was stated that ten main line diesels – D33 to D42 – would be allocated to Bristol Bath Road, not Barrow Road, for working on the route and on 10 April Type 4 'Peak' D93 arrived in Bristol for crew training. Some express passenger workings like the up 'Devonian' were diagrammed for diesel haulage from the start of the summer timetable, 12 June, with diesels gradually taking over more trains. From 6 November diesels also commenced work on some of the most important fast fitted freights, such as the 5.50am Washwood Heath-Westerleigh and the 4.48pm Bristol-Leeds.

Attention during the year is focussed on the Bromford Bridge oil tank trains worked from Avonmouth and Fawley. This was important traffic for BR, but reports reveal that, even using the most modern steam power, 9F 2-10-0s, problems often arose, leading to big delays which reflected poorly on BR's ability to deliver the goods on time.

JANUARY TO MARCH 1961

'Although this engine worked the train from Sheffield it did not go to shed there and was previously serviced at York. I am therefore drawing the attention of the DMPS York to this failure in view of the condition of the tubes when the engine came off at Gloucester.'

Driver F E Ford had an early failure to report, on 3 January. Along with his fireman, T Phelps, he worked the 6.30pm Birmingham slow from Bristol with 44226 and had to report 37 minutes lost due to 'engine not steaming, small tubes blocked up firebox end. Engine changed at Gloucester.' The Barnwood foreman wrote 'Fire, ashpan and brickarch in good condition. Coal appeared to be of good quality but rather small.' The engine's home shed, Saltley, sent servicing details – tubeplate last cleaned 31 December, brickarch cleaned 7 December. So nothing very dramatic, but enough to delay the train.

In times past, this train was often a Compound working – the author rode behind Derby's 40925 on the working on 31 March 1959. But 4Fs were good on these stopping passengers, their small wheels giving rapid acceleration away from the frequent station stops.

Another Barnwood driver had a different Saltley 4F, 44171, from Bristol on 6 January, with the 7.43pm class 'E' freight to Bescot, routed via Worcester and Stourbridge. But it came off the train at Gloucester due to having insufficient coal for the job. The Barnwood shed master wrote 'When questioned as to his reason for leaving Barrow Road Loco with the tender obviously insufficiently coaled, Driver Greatorex replied that the engine

had been prepared for him and left in the sidings, and it was not until after leaving Bristol that he became aware that the coal was so low.'

On 7 January 45418 from another Birmingham shed – 21D Aston – had the 5.0pm Bristol-York express, worked by Barrow Road men. It became a casualty at Gloucester short of steam due to a number of centre row firebars collapsing, filling the ashpan with fire. 22 firebars in total needed replacement. 73003 worked the train forward.

The same train was in trouble again on 10 January, with Derby's 45612 *Jamaica* at the head. Again short of steam, this time due to a portion of the brickarch collapsing, 45612 came off at Engine Shed Junction Gloucester, presumably to facilitate an engine change from the adjacent Barnwood shed. Delay was 16 minutes.

The next day, the 7.25pm Bristol-Newcastle mail also stopped for attention at Engine Shed Junction, whilst a new flexible carriage warming pipe was fitted to 45683 *Hogue*, so that the train could be steam-heated. Smart work here, as the delay was just eight minutes.

'Jubilees' from 8A, Edge Hill, Liverpool, were not common at Gloucester, so the appearance of 45554 *Ontario* on 16 January was an out-of-course turn. It had worked the 12.48pm express from York throughout, but was short of steam on arrival at Gloucester, caused by a dirty fire and very small coal. Possibly 45554 had been 'borrowed' by York depot at short notice to work this train, instead of heading back home, probably on an express over the Standedge route.

According to former Eastern Region shed master Eric Beavor, 'York was probably the most notorious depot in North-East England for using any strange engine to cover ... their workings.'

Also on 16 January, 73003, mentioned a few days ago as replacing a failed loco, was itself having trouble – keeping the guard warm on the 8.15pm Bristol-Leeds parcels! The casualty report states 'Guard refused to work forward owing to his van not being steam heated.' Quite understandable in the middle of January! The intermediate carriage warming pipe between engine and tender had burst. A delay of 44 minutes ensued at Gloucester while 43985 was rustled up to work the train forward, no doubt with a very warm guard.

Two Barrow Road engines were short of steam in the early hours of 18 January. Barnwood's Driver Bill Dipple with Fireman A Green were on the footplate of 45682 *Trafalgar* working the previous night's 9.5pm Bradford-Bristol mail forward from Derby. The loco struggled, losing 29 minutes to Birmingham, where it was replaced by 45280. A few hours later, 45504 *Royal Signals* pulled out of Washwood Heath yard on the 5.50am express freight to Bristol with a load of 40 on, but gave up at Gloucester, with a 13 minute delay.

Hot boxes on 4Fs were not uncommon, especially it appears when working passenger trains or fast freights. 44552 was on the

4.20pm Burton-Bristol class 'C' beer train on 20 January. Barnwood Driver J Daniels with Fireman M Scadding reported 'On arriving in Gloucester Yard I discovered that the right leading axle of engine was quite hot and not fit to go forward to Bristol.' The cause was shown as 'Axlebox back-pressure valve top cap-nut loose.' and repairs were effected at Barnwood. 45659 *Drake* took the beer to Bristol.

Saltley 'Black Fives' were such an everyday sight at Gloucester, that it took a while for someone to realise that 44669 which failed on 23 January working the previous evening's 6.28pm Sundays only Newcastle-Bristol passenger, M319, was actually based not at 21A Saltley, but 12A Carlisle Kingmoor. So reports of its last washout, exam etc were a bit delayed in being received from Kingmoor, the request for this information having initially been directed to Saltley. The loco gave up at Gloucester as the driver stated 'Exhaust injector not working and live steam injector would not maintain boiler, also short of steam.' It had possibly worked the train some distance, perhaps all the way, as the fire was in a very dirty condition, there was a heavy accumulation of clinker on the firebars and most of the coal left was slack. 45268

– of 21A – worked to Bristol.

As it happened, the following express in the early hours of 23 January, the 9.5pm Bradford-Bristol, train M277, suffered a slight delay, 10 minutes, due to the tender brakes binding on 21A's 45040; it appears to have worked on through to Bristol.

Timekeeping of the Esso block trains from Fawley to Bromford Bridge was a cause for concern to top Western Region management, with the Assistant General Manager at Paddington, no less, keeping an eye on things. On 23 January, 92231, allocated that month to 71A Eastleigh especially for the traffic (along with 92205 and 92206), delayed the 12.25am ex-Fawley working by 167 minutes. The Barnwood driver's report is particularly informative, so is reproduced in full.

44226, which had failed on 3 January, did so again at 1.3am on 25 January. It was working another fast goods, the 8.55pm Avonmouth-Water Orton with 20 loads. It came off at Engine Shed Junction, where the train was due at 11.14pm. Fifty minutes of the lateness was booked against the engine, which had a dirty tubeplate and blocked tubes to blame for the casualty.

BRITISH RAILWAYS (WESTERN REGION)

B.R. 32852

LOCOMOTIVE CASUALTIES—NON-MECHANICAL

Station. **Gloucester Barnwood**   District. **Gloucester**   Date. **24.1.61**

Report of failure of Engine No. **92231**   Class. **9F**   Home Station. **Eastleigh**
to complete scheduled working.

Driver : **G. Cornwell, 85C.**

Date of Failure **23.1.61**	Nature of Failure **Short of steam, insufficient coal.**
Time of Failure **en route**	
Locality of Failure **Cheltenham**	Condition of Ashpan after failure **Fair**
Train :— **11/20pm.**	Condition of Fire after failure **Very dirty**
	Condition of Tubes after failure **Clean**
Time **12/25pm**	Amount of coal left on tender **6 cwts**
From **Fawley**	Whether fitted with S. Cleaning Smokebox
To **Bromford Bridge**	Report on condition of tubes after failure to be completed by
Pass. or Goods	boilersmith **Tubes in good condition.**
Booked Load	**Hole in brickarch.**
Actual Load	Foreman's Remarks **Fire very dirty, irony clinker,**
State of Weather **Fire**	**poor quality coal left on tender.**
Delay : 167 minutes	

Servicing details before leaving starting points

Engine worked from **Eastleigh**	Actual time engine left shed **Eastleigh to say**
Actual time engine arrived on shed **E'leigh to say**	Date Tubeplate Cleaned **"**
Date Tubes Cleaned * Rodded or Blown **"**	Date Brickarch Cleaned **"**
Whether fire cleaned or completely dropped **"**	Last Washout **"**
Ashpan and Smokebox cleaned **"**	

Coal supplied Amount	Colliery and Grade	Colliery and Grade	Has difficulty been experienced with this grade of coal ?
	Eastleigh to say		

*Delete as necessary.

Details of Reports Regarding Bad Steaming of Engine During Six Previous Trips

Eastleigh to say.

DRIVER'S/MOTORMAN'S REPORT

B.R. 32841

BRITISH RAILWAYS

Glos. (85C) Depot .... Region. Date Mon. 23rd Jan 19 61

Place of Incident ...... Cheltenham

Date of Incident 23 Jan 19 61

Driver Motorman } * G.Cornwell Reg. No. 102 Depot 85C

Fireman Second Man } * J.Claridge Reg. No. 241 Depot 85C

Guard F.Whitmore Station Glos.

11/20 p.m. Pgr. } * Train from Fawley to Bromford

12.25 a.m. Frt. } Train from Fawley Actual load 26 tanks & 5 runners Assisting

on .... 19 ...... Reg. load ...... Train

Locomotive/Diesel/Power Car Electric Multiple/Diesel Elec./Unit. } * No. 92231 9F

Class/Type*

*Delete words not applicable

Failure of engine at Cheltenham.

To commence with, I am given to understand that this engine worked through from Fawley instead of being changed at Westbury in accordance with the diagram. The result was that the fire was in a very dirty condition and the Bristol driver whom I relieved stated that they had had some difficulty in raising the required amount of steam for the job. This state of affairs became hopeless after leaving Gloster E.S. Jct and I decided in the interest of everything concerned to bring the engine back from Cheltenham to Gloster Loco to be reconditioned. I might add that the coal which did not appear to be of very good quality, was also getting scarce and right at the back of the tender. To make matters worse, the water sprinkler apparatus on tender would not shut off properly with the result that my mate was getting wetter and wetter every time he made an effort to get coal forward. I should like to add that I do not consider that the coal capacity of this type of tender is sufficient for a heavy train of this description over a long journey unless coaled to capacity with Welsh coal.

Signature G.N.Cornwell.

Around one and a half hours later, at 2.35am, another loco casualty arrived at Engine Shed Junction in the shape of Saltley's 92139 with the previous day's 2.25pm Fawley-Bromford Bridge Esso working. This train was due to pass Engine Shed at 10.16pm, though only 62 minutes of the lateness was booked to the engine for this failure. Lack of coal was quoted as in the last report for this train. The engine had worked through from Fawley. Its BR1C-type tender could hold nine tons of coal, against seven tons for the BR1G tenders of the three 9Fs allocated to Eastleigh, but still seems to have been insufficient for the job.

A Burton 8F, 48694, found itself on the 12.25pm Fawley-Bromford Bridge on 27 January. Having worked through from the south, it came off the train at Cheltenham at 8.0am. The Barnwood foreman was none too happy with his driver, stating that had attention been given to the thin irony clinker covering the bars, the delay of 44 minutes could have been overcome and the engine worked forward. But Driver Daniels said that the crew started to clean the fire, but the tools were too short. And he had then been informed that Control was sending another engine from Gloucester to take over.

44962 suffered a mechanical failure on 3 February when working the 10.30am Bristol-Newcastle express. The combination lever and union link of the left side valve gear broke and fell away. 45725 Repulse was available at Gloucester to take this heavy train, 376 tons, northwards, with just a 14 minute delay. The casualty was sufficiently serious for Mechanical Inspector A E Humphries to be sent from Birmingham to examine the loco at Barnwood and compile a report, which is reproduced. New parts were ordered from Derby Works but it was 14 March before 44962 was available for further work.

Next day, Saltley's 43507 was working the 11.20am class 'H' goods from Bristol to Washwood Heath, but became a casualty at Cheltenham with the right big end hot, metal fused and brasses broken. The train went forward with classmate 43309 after a 68 minute delay. 43507's last shop repair had been at Derby as long ago as 29 January 1954, since when its estimated mileage was 75,650. Saltley stated that the 3F's last mileage exam had been on 3 August 1960, since when it had only done an estimated 150 miles, so it wasn't getting much use. Before setting off from Barrow Road shed that morning, a fitter had done some work on the left hand cotter, but there was no mention of the right hand side. Management's conclusion was that the high mileage since 43507's last shop repair was a contributory factor to the failure.

Later on 4 February 17A Derby's 45557 New Brunswick failed short of steam on the 12.43pm Newcastle-Bristol express, taken forward from Eastgate by 73016. The cause of the casualty was shown as 'Large hole in brickarch, tubeplate very dirty.' As the Barnwood shed master wrote 'This engine failed with the same defect on 12 January whilst working the 10.25am 'A' Manchester-Bournemouth.' Perhaps Barnwood felt disinclined to put right another shed's failure and sent 45557 back light engine to its home depot for attention. (45557 was having a bad run – on 7 January, it derailed in Bath Green Park station.)

Not much more than 24 hours later, another Derby 'Jubilee'

Left: Saltley's 44962 appears to be in good shape at Barnwood shed, but it spent well over a month here after failing on 3 February 1961, while hauling the 10.30am Newcastle express.

To: D. L. Pride Esq.,   From: Mechanical Inspector
     D.T.S. (R. & M.) BIRMINGHAM   D.T.S.O. BIRMINGHAM.

Engine 44962 at Gloucester

Sir,

I visited Gloucester to examine the above Engine which became defective while hauling the 10.30 Bristol – Newcastle train on Friday 3/2/61.

From enquiries I made, it was first noticed that something had gone wrong at Westerleigh but the L.H.D. piston valve centralised itself and the train reached Gloucester.

The broken combining lever and union link was recovered near Frocester approximately 12 miles away.

The L.H. piston valve had been removed when I arrived and proved to be in a very dry condition with a small portion of the front head broken away on the exhaust side allowing the first ring to brake up various fragments of which were recovered from the cylinder cocks. The valve sleeve itself was very dry and varied in size from 10.085–10.131.

It was evident from the amount of oil which had flowed down the outside of the cylinder wrapper from a leaking connection that very little had gone to the valve sleeve.

The combining lever had broken off about 2½ inches below the valve crosshead pin hole and after thrashing about had finally broken the union link and fell to track. On examining the top part of the combining lever it showed no sign of crystallising and was of smooth fine grain.   Other damage caused was as follows:-

L.H. Piston rod scored
Oiling ring to piston gland broken
Castle nut of gudgon pin damaged
Outside portion of valve crosshead guide distorted

It would appear that the prime cause of failure was the partial breakdown of the lubrication system throwing undue strain on the combining lever which after breaking free thrashed about between the piston rod and the top slide bar is evidenced by marks on main frame and finally broke the unionlink.

All pins and bushes of valve gear are in good condition.

After blanking off atomisers oil was finally obtained in valve sleeves.

I have arranged for the piston and crosshead together with the piston valve, crosshead and guides to be sent to works for repair. Repairs to lubrication system to be carried out at Gloucester.

                                  (Signed A. E. Humphries)

did not get far before failing. Gloucester Driver E V Smith and Fireman L Evans were waiting at Derby station with 45579 *Punjab* on 6 February to take over the previous night's 9.5pm Sundays Bradford-Bristol class 'A', which arrived from Sheffield behind 41B Grimesthorpe 'Crab' 42794, substituting for the power rostered to work through from Leeds to Bristol, a Barrow Road 'Jubilee'. The train was due away from Derby at 12.25am, but did not arrive until 1.20am, so perhaps Barrow Road's engine had failed en route from Leeds. After changing engines, the train left at 1.32am, perhaps with the crew hoping to make up some time. But out on the road, 45579 started priming very badly, caused by being overdue for a boiler washout. The driver reported '...you could not work the engine with one inch of water in the gauge glass with regulator shut without the boiler priming very bad...' Anyway it struggled to Duddeston, where 73002 of 41C Millhouses came from the nearby Saltley depot to take over, with a 44 minute delay. Even then, things did not go smoothly, five minutes were lost due to the engine slipping in the tunnels between Birmingham New Street and Church Road because the sanders were not working. On the face of it, the crew would have been better off keeping 42794! 'Crabs' were regularly seen on stopping passenger trains through Gloucester, but their class 'A' work was normally limited to excursions and reliefs by this date, with one known exception – the 1.10am Bristol-Sheffield mail train was diagrammed for a Saltley 'Crab' on Mondays only in the timetable in force from November 1959.

92007 managed to fail twice on 6 February. It started from Avonmouth on the 11.0am Bromford Bridge tanks, load 25 equivalent to 86 wagons, but came off at Gloucester. Just a small defect, but to a vital part, the whistle cable was defective, so 48037 worked the train from Barnwood, after a 50 minute delay. The defect was put right and the engine was then given the 9.45pm fitted freight to Water Orton. But there was a problem with the vacuum when the loco coupled to the train, which could not be put right without returning to the depot. So 43985 came off shed to do the turn, with a 45 minute delay.

The very next day, the 11.0am Avonmouth-Bromford Bridge was in trouble again, this time 92248 was short of steam. Barnwood Driver L M Davis and Fireman Ken Huggins took over from a Barrow Road crew at Gloucester South. As Driver Davis stated '...was told engine was not steaming, found brick arch in fire, stopped at Cheltenham High Street to remove same from 2.0pm to 2.30pm, worked forward to Bromford Bridge.' The DMPS expressed his concern at the delay 'in view of the importance given to this traffic.'

That wasn't the end of the troubles for the same train and same crew that week. On 10 February 92221, again a Barrow Road engine, lost 30 minutes short of steam on the working. Driver Davis reported 'I relieved Bristol men at Gloucester Engine Shed, was told engine not steaming, large and small tubes stopped up, tube plate dirty, bad, stopped at Spetchley from 3.0 to 3.15pm short of steam, also at Bromsgrove South from 3.50 to 4.5pm.' The delays made the train exactly two hours late leaving Bromsgrove South.

One of the other block oil trains, the 2.25pm from Fawley with 30 tankers, had problems on 6 February. The Gloucester driver said a dirty fire on the loco, signal checks at Pirton and Stoke

Works, plus a gale force wind all contributed to the 29 minute loss. The only thing was, he said the engine was 92005, while Eastleigh shed said the train had been worked by 48647. In fact 48647 worked the 11.20pm ex-Fawley on the day in question, and after investigation by Barnwood, everyone finally agreed – on 12 May, three months after the problem! – that the loco had actually been 92205, one of the Eastleigh allocated 9Fs.

The 2.25pm Fawley lost 15 minutes on 9 February, with the same crew, Driver H Poyner and Fireman C Folkes, due to a dirty fire and the front half of the brickarch missing on 15A Wellingborough's 9F 92118.

Barnwood's 44045 had a boiler washout on 10 February, as well as having the tubes, tubeplate and brickarch cleaned. After this attention, it was ready for that evening's 8.55pm Gloucester to Bristol class 'H' freight. The job done on 44045 was too good, tempting other depots to borrow it for their work. Barrow Road seems to be chief suspect, as by 16 February, it was on that depot's turn, the 2.45am Washwood Heath-Westerleigh class 'C' express fitted freight, actual load 43, with Barrow Road Driver R J Lacey at the regulator from Gloucester. It completed the job, but lost 20 minutes, suffering shortage of steam due to a dirty fire. The ashpan was full on arrival at Barrow Road, so maybe it had not been getting the attention lavished on it by Barnwood. 44045 did not get back home until 22 February, arriving off the 7.30am class 'H' freight from Washwood Heath. The LMS instituted a system from the 1930s whereby Barrow Road should have informed Barnwood of the trips which had been worked by 44045 while it was away. This was to enable the home depot to calculate when the loco next required a boiler washout, which would normally be done after a certain numbers of days use.

Although Barnwood men apparently did not work on the 'WD' 2-8-0s allocated there (see reports from November 1959), they did not miss out completely on this dubious privilege. Driver D Powell and Fireman Stinchcombe had 90346 of 2F Woodford Halse on 16 February with the 7.15am freight from Honeybourne to Severn Tunnel Junction. They had worked up from Gloucester to Honeybourne with no problems, but after passing Winchcombe on the return, both injectors failed. They put the train off at Cheltenham Malvern Road and went to shed there. The driver noted 'Examination by myself and the fitter revealed that there was no supply of water to the injectors running from the tank, although there was three-quarters of a tank full of water.' Subsequently it was found that the sieve in the tender tank was blocked with soft sludge. Once this was removed the injectors worked without problem.

Hardly a week had passed before another delay arose with the 2.25pm Fawley-Bromford Bridge tanks. 92206 lost 30 minutes on 17 February after leaving Gloucester, but Driver G N Cornwell was determined not to take the blame for this. His report is copied here. Note the 'dig' at 'ex-GWR officials' – showing the Midland/Western divide was still very much alive in the area, well over a decade after nationalisation!

The 11.20pm Fawley-Bromford Bridge arrived at Gloucester in the early morning of 22 February behind 18A Toton's 92153 which had worked through. But Barnwood Driver H Burrows was told by the Bristol driver he relieved that the engine was not steaming well. Having set off, things went from bad to

DRIVER'S/MOTORMAN'S REPORT

BRITISH RAILWAYS

B.R. 32841

Gloucester (85C) .... Depot ...... Western ...... Region. Date .... 23rd Feb. 19 61

Date of Incident .... Fri. 17th Feb. 19 61 .... Place of Incident ....

Driver/Motorman }* G.N.Cornwell .... Reg. No. .... 102 .... Depot .... Glos.(85C)

Fireman/Second Man }* J.Claridge .... Reg. No. .... 241 .... Depot .... "

Guard .... Station .... Saltley

2/25 ~~xxxRxxx~~ }* Train from .... Fawley .... to .... Bromford
p.m. Frt.

on (16) 17.2. 19 61 .... Reg. load .... Actual load ....

Locomotive/Diesel/Power Car Electric Multiple/Diesel Elec./Unit. }* No. 92206 .... Train .... Assisting
9F

Class/Type*

*Delete words not applicable

Sir,

I wish to report that I was taken off my job to relieve the above train which I did at 12.35 at Gloster E.S.Jct and left there at 12.40 after the train had been standing a considerable time awaiting relief. I had my doubts about having sufficient coal on the tender, but bearing in mind the delay already suffered by this train, I set off on my journey and by careful manipulation, although not keeping strict timing, I managed to reach Duddeston Rd where I was relieved. There was now about 15 shovefuls of slack coal left on the tender and the engine had to detach and proceed into Saltley Loco to obtain sufficient coal to finish the job, this causing about another 30 mins delay. I might add that, although these trains are diagrammed to change engines at Westbury, this one had worked through ~~xxxxx~~ and this is the second time I have had this experience of shortage of coal. This state of affairs is taking place regularly and in my opinion is bringing much

Signature .... contd...

discredit on us ex L.M. staff who work these trains from Bristol and for which we are in no way to blame and leaves me with only the conclusion, with all due respect, that some ex G.W.R. official in the Bristol area either does not care, or is out to bring about this discredit, but will only suceed in losing this traffic before very long and get no benefit for doing so.

(Signed) G.N.Cornwell.

worse by the time the train got to Cheltenham High Street at 7.30am. They went into the loop there, as the fireman was unable to maintain steam or water. The fire was 'very bad' and there was a layer of hard clinker on the bars. After talking to Control, the engine went back light to Barnwood, where another Toton 9F, 92158, had been prepared for the job. Delay was put as 112 minutes. The DMPS, wrote in his conclusion to the report 'I am asking the DMPS, Eastleigh to note this case and ensure that in future special attention is given to this engine working.'

Just over 36 hours later, the 2.25pm Fawley was on report – again – due to 92206 having insufficient coal to complete the working. The train arrived at Engine Shed Junction at 11.35pm, the loco went to Barnwood for coal, and the train was away at 11.59pm, nearly an hour and three-quarters late.

Gloucester Docks shunter 41537 emerged from Derby Works after overhaul and was making a fairly leisurely return home as befits a small engine, but it nevertheless managed to run hot axleboxes. It was seen on 20 February at Saltley shed before being towed through Cheltenham a few days later and on 26 February was awaiting attention at Barnwood.

Reports from some years earlier, 1951, indicate that 41537 was in poor condition and was thought to be due for scrap, but it went to Derby works and got an overhaul instead.

A visit to Barnwood depot on 26 February found 33 steam locos and one diesel shunter, including a notable 'foreign' engine – O1 class 63589 (The definition of a foreign engine is one on hand at a point other than that to which it is allocated). At least six of the 33 steam were casualties under repair and would be listed on the shed's daily return of 'Locomotives stopped, under or awaiting repair' – they were: 6879; 41537; 44551; 44962; 45668; 92241. 41537's driving wheels and axleboxes were sent to Derby for repair and the loco returned to service on 13 March. 44551 was a 17B Burton engine, which had run hot around 20 February, probably on the Burton beer, and it was attended to at Barnwood; its shedmate 44552 had also run hot with

Above: Two Deeley 0-4-0Ts were at Gloucester for years to shunt in the Docks area. One of them, 41537, emerged ex-works from Derby in February 1961, but sustained hot axle boxes while being towed back dead to Gloucester. It is pictured on that journey in the down loop outside Cheltenham Lansdown station coupled to a 3F. There is an oil can on the tank top, perhaps indicating oil has been poured into the boxes, but the damage was probably already done. 41537 required attention at Barnwood before it could resume duties.   *R Wales*

the beer train on 20 January as previously noted. 6879, 45668 and 92241 were further hot box casualties.

Having suitably chastised the Southern Region recently, the DMPS turned to the Eastern Region which got into his bad books when 55A Holbeck's 45573 *Newfoundland* failed at Gloucester on 28 February with the 4.45pm Bradford-Bristol express loaded to 11 coaches, 360 tons. The loco was short of steam, with both large and small boiler tubes blocked. The DMPS reported 'Although this engine worked the train from Sheffield it did not go to shed there and was previously serviced at York. I am therefore drawing the attention of the DMPS York to this failure in view of the condition of the tubes when the engine came off at Gloucester.'

The same train lost 15 minutes on 14 March, when Millhouses's 45602 *British Honduras* was at the front end. Barrow Road driver Basil Curtis stated: 'This engine was priming very bad, it couldn't reach 40 miles per hour between Ashchurch and Cheltenham, could not use second regulator. Engine wired off at Gloucester.' Apparently the injectors were also playing up. Its last boiler washout had been on 28 February and it was also treated with X-Zit powders (placed in the fire to help clear the tubes) while on Holbeck shed that day. 73016 took the train on to Bristol.

Barrow Road's own engines were ailing later in the month. Patriot 45506 *The Royal Pioneer Corps* had the 2.15pm Bristol-York express on 20 March. It only got as far as Gloucester before giving up short of steam and with a defective exhaust injector. 45598 *Basutoland* took over the train and 45506 went back light engine to its home shed. There it was found that the failure was caused by the smokebox door drawing air. The loco had covered 4000 miles since an overhaul at Crewe Works in January.

45685 *Barfleur* from Barrow Road depot had the 12-coach 9.2pm Bradford-Bristol express in the early hours of Monday 27 March, but according to Barnwood Driver F Monk, it 'primed all the way from Derby to Birmingham.' 30 minutes was lost to New

Street, where shedmate 45651 *Shovell* took over.

The next day, 82E's 45519 *Lady Godiva* had the 5.50am class 'D' express freight from Washwood Heath to Westerleigh. It was short of steam when it was failed at Gloucester, the delay being one hour and thirty minutes. After examination at Barnwood by Foreman M V Smith, he found 'Fire, smokebox, ashpan and tubes were clean and brickarch in good condition.' It was concluded that poor quality coal was the culprit this time and the DMPS at Saltley, where the loco had been coaled, was informed accordingly.

Trains Illustrated noted that most duties at the time for Bristol's Patriots were parcels and freight, though both 45504, just back from Crewe Works – with a 12B Carlisle Upperby shedplate! – and 45506 resumed work on express passenger trains in April.

The 5.47pm Birmingham-Bristol stopping passenger had problems on 28 March after changing engine – and crew – at Gloucester. 44818 came off and Barrow Road's 75004 took the train of 12 vehicles, 261 tons, away from Eastgate. The engine got short of steam, stopping in section beyond Coaley Junction and again at Charfield for a 'blow-up'. Eventual arrival at Temple Meads was 10.26pm, 37 minutes late, of which 20 minutes was down to the engine.

APRIL – JUNE 1961

'Boiler examined by Derby boiler inspector on 25.5.61 who found that the repairs required were too heavy to be undertaken at the depot. The engine is to receive factory attention before being put back into traffic.'

The 1.10am Bristol-Sheffield mail was in trouble on 1 April, with 'Jubilee' 45626 *Seychelles* succumbing at Ashchurch due to the failure of both injectors. Barnwood Driver W Powner

OTHER CASUALTY REPORTS January-March 1961

Date	Loco	Home shed	Train	Problem	Place	Delay (mins)	Notes
04/01/61	44211	21A	11.15am WWH-Bristol class C	HOT	GLOS	55	1
07/01/61	44818	21A	10.15am Bradford-Paignton class A	INJ	GLOS	7	2
12/01/61	45557	17A	10.25am Manchester-Bournemouth class A	SOS	ER	5	3
18/01/61	45607	41C	10.45am Sheffield-Bristol class A	CWA	ER		4
23/01/61	73135	17A	8.30am Cardiff-Newcastle class A	SOS	STOKE	12	
25/01/61	45342	17A	6.18am Derby-Bristol class A	INJ	ER	32	5
27/01/61	73136	17A	9.2pm Bradford-Bristol class A	SOS	ER	12	6
30/01/61	45668	17A	7.35am Nottingham-Bristol class A	HOT	GLOS	30	7
30/01/61	45577	82E	12.48pm York-Bristol class A	SOS	ER	8	8
07/02/61	45649	17A	6.18am Derby-Bristol class A	INJ	ER	5	9
21/02/61	75009	85C	5.15pm Bristol-Birmingham class B	MECH			10
23/02/61	45682	82E	7.40am Bristol-Bradford class A	SOS	ER	3	11
25/02/61	44853	55A	9.2pm Bradford-Bristol class A	SAND	BNS	7	
26/02/61	45699	82E	1.20pm Bristol-York class A	SOS	ER	4	12
28/02/61	73046	41C	8.40am Bristol-Sheffield class A	CWA	ER	13	13
08/03/61	42791	21A	1.0pm WWH-Gloucester freight	MECH		30	14
10/03/61	92152	21A	11.40pm Avonmouth-Water Orton class D	MECH	GLOS	74	15
21/03/61	45564	55A	8.30am Cardiff-Newcastle class A	SOS	ER	15	
29/03/61	45610	17A	8.30am Cardiff-Newcastle class A	PRI	ER	11	
31/03/61	44662	55A	8.0am Bristol-York class A	VBI	ER	10	16

Notes

1	44211 repaired at Barnwood. 75022 took train onwards
2	44818 worked through to Bristol
3	45006 worked train from Gloucester. 45557 returned to home depot
4	45602 worked train from Gloucester. After repair 45607 worked 4.5pm Bristol stopper
5	44853 took train from Gloucester. 45342 returned light engine to home depot
6	44804 took over train at New Street. 73136 to Saltley for attention
7	45253 took train forward. 45668 repaired at Barnwood
8	42839 took over at Gloucester. 45577 sent to Barrow Road for repairs
9	44661 worked train from Gloucester. 45649 repaired at Barnwood
10	Loco failed at Lawrence Hill. No further details
11	73155 took train on from Gloucester. 45682 returned light to Barrow Road for water test
12	45699 off at Saltley for attention. 45447 took over
13	45683 took train from Gloucester. 73046 repaired at Barnwood
14	Boiler pressure gauge failed at Bromsgrove, new one fitted and loco worked on
15	Blower pipe fractured, repaired at Gloucester. 92225 took train north.
16	Repairs effected at Worcester, loco worked onwards

reported the train was delayed there from 3.1am to 4.20am and that 'the fire was about to be disposed of to guard engine boiler when live steam injector was induced to operate.' He went on to say 'Subsequent time lost due to boiler priming and tender-first running of (replacement) engine 42761.' Meanwhile 45626 went back to Barnwood where no defect was found with the exhaust injector, while the regulating valve of the live steam injector was tight in the body and was freed up to work without a problem. 42761 worked down through Cheltenham on a freight later that day, while 45626 was seen back in action on 3 April.

Classmate 45654 *Hood* had the 1.10am on 3 April with 15 vehicles, slightly over the maximum load of 420 tons, and stopped short of steam at Pirton sidings. Driver F E Ford, with Fireman Ken Huggins, reported 'Train stopped for steam and water caused by dirty fire and tubes, also brickarch fell into fire.' 26 minutes were lost on the journey to Birmingham. The brickarch had last been examined at 45654's home shed of Millhouses on 22 March, also the date of its last 'X' exam.

On 4 April Barrow Road's 73003 had the 7.40am Bristol-Bradford express – at least it did until failing at Gloucester short of steam. Apart from a dirty fire, the smokebox door was warped and drawing air. Barnwood reset the door, cleaned the fire and sent the loco back to its home depot with a repair card. The report complained the loco had not been serviced properly at Barrow Road. 73003's last works visit was to Swindon, receiving a heavy intermediate overhaul dated 12 November 1959, since when its estimated mileage was 63,711. The replacement loco on this occasion was well away from its home depot, being 14A Cricklewood's 44777. Having found itself on the Bristol-Birmingham route, 44777 appears to have been 'borrowed' for quite a number of turns in the area, as the author saw it – complete with 14A shedplate – at Cheltenham on 10 April. (As with quite a number of previously 'rare' engines, 44777 gravitated to Saltley in later years, becoming an everyday sight.)

Gloucester's opinion of Barrow Road's servicing of engines was probably coloured by the fact that another loco they turned out later that day also failed short of steam at Gloucester. 73016 of Millhouses had 1E69, the 9.15am Paignton-Leeds relief to the up 'Devonian'. When examined at Barnwood, Foreman E Mills wrote 'Fire in poor condition, with several large and small tubes blocked firebox end.' The loco had been on Barrow Road from 12.40am until 12pm when it departed to take up this duty. Its tubes had last been cleaned at Millhouses on 11 March.

However Barrow Road was soon in a position to make the same criticism of Barnwood, after problems with the down 'Devonian' that same day, April 4. Leeds Holbeck's 45564 *New South Wales* was short of steam on arrival at Eastgate, due to a very dirty fire and partial collapse of the brick arch. Even so, only 10 minutes delay was booked against the engine. The standby engine at Eastgate, Saltley's 44962, took over, but was also short of steam, losing an unspecified amount of time on the journey to Bristol, causing Barrow Road to implement a casualty report. In response, the Barnwood shed master reported 'This engine arrived on this depot at 7.15am on 4.4.61 off the 3.55am (freight 5V10) ex Birmingham and no repairs were booked against it. Control were diagrammed to provide a Class 5 engine for 2.0pm standby on this day and engine 44962 was turned off shed 1.45pm

for this purpose.' During its sojourn at Barnwood, the fire, ashpan and smokebox were reported as cleaned.

As an interesting comparison with 73003's estimated mileage of 63,711 mentioned earlier, 45564's mileage since an intermediate overhaul at Crewe dated 26 November 1959 had been 81,260. So with very similar shopping dates, the 'Jubilee' had covered around 17,500 more miles than the 'Standard'.

Another loco casualty later that day saw 43017 arrive with Driver J Workman at Tewkesbury shed off the 5.10pm Birmingham-Evesham-Ashchurch passenger with defective vacuum. It was deemed unsuitable to work the following morning's 7.14am Ashchurch-Birmingham passenger as diagrammed, this information being conveyed to Barnwood by Chargeman J Page. The shed master there stated it was not possible to effect a change of engine that evening as Tewkesbury shed closed at 8.15pm. He continued 'Arrangements were, therefore, made for a suitable engine to be booked on the 4.42am (Barnwood-Bromsgrove) 'Pick Up' on 5.4.61 and for engine 43017 to run light Tewkesbury to Ashchurch by Driver J Broughton who would normally have taken it off shed for the 7.14am passenger, and change engines at Ashchurch.' This was done; unfortunately there is no mention of which 'suitable engine' worked the pick up from Barnwood to Ashchurch and then the passenger to New Street. The 4.42am would usually be a 3F 0-6-0, which, of course, was perfectly capable of taking the passenger, but as the 7.14am – Saltley engine turn no.26 – called for a class 4 tank, perhaps something bigger was provided on this occasion. The Tewkesbury crew were rostered – on their turn no.5 – for the passenger as far as Evesham, where Saltley men, on turn 145, took over.

Meanwhile, 43017 hauled the Bromsgrove pick up with Driver D Davis and Fireman T Etheridge of Barnwood. They reported 'After we had left Ashchurch the vacuum went altogether. So when we got to Bromsgrove and put our train off we took the engine onto Bromsgrove shed to have it put right, which the fitter did.' The problem was found to have been caused by a sponge cloth wedged in the train pipe under the ejector box. The engine off this working was also booked to do a trip freight up the Lickey incline to Blackwell at 9.45am – perhaps another loco did it this day while 43017 got attention.

Later on 5 April, Barrow Road's 45690 *Leander* worked the 12.48pm York-Bristol express, but became a casualty en route due to priming and a defective brickarch. Delay was minimal though, just five minutes, with 45564 *New South Wales* shown as taking over at Gloucester. Now 45564 had failed the day before as noted earlier, and, according to that report, was returned to its home depot, Leeds Holbeck, for attention, where a new brickarch was fitted. Presumably this was after it had worked this train to Bristol. Anyway, 45690 was also sent to its home depot for attention; it was not out of action long, being seen on an up express on 11 April.

A mileage comparison between 45690 and 45654, mentioned on 3 April, is interesting as both were outshopped from Crewe around the same time. 45690 had a heavy general overhaul dated 6 May 1960 with an estimated mileage since of 44,712. 45654's estimated mileage since an intermediate overhaul dated 26 May 1960 was 45,500.

The 1.20am Barnwood-Washwood Heath class 'H' had unusual power on 6 April in the shape of K3 61894 of 40E Colwick; it had worked into South Wales the previous day on a special freight.

Another failure from 4 April, 73016, was called into action on 6 April, to take over the previous night's 11.20pm Fawley-Bromford Bridge tanks. Barnwood Driver L E Dilleigh and Fireman T Phelps crewed Eastleigh's 92205 on this train, working up from Bristol, when they noticed three washout plugs blowing. The train was delayed for 67 minutes changing engines at Gloucester. Examination at Barnwood depot found no less than six washout plugs blowing! Eastleigh reported the engine had been stopped on 10.2.61 for boiler repairs and piston and valve exam. The boiler washout date was 27.2.61, with a steam test not until 30.3.61 when one plug was removed and a hole tapped. It had done a turn of duty on 4 April, on the 2.25pm Fawley-Bromford Bridge. Eastleigh commented 'It is difficult to know why the plugs were leaking.' Gloucester requested the DMPS at Eastleigh to follow up suitably with the staff concerned, stating 'Apparently the boiler washout plugs had not been properly tightened after washing out and steam testing'. The loco was new in April 1959, with an estimated mileage since of 61,000; a lot less on an annual basis than the two 'Jubilees' mentioned a little earlier.

Despite fitter's attention to the sanders of 45447 on both 6 and 7 April at Barrow Road, the loco stalled coming out of Birmingham New Street on 10 April while hauling the down *'Pines Express'*. Driver R Wheat, with Fireman J Nicholls, of Barnwood reported 'The train came to a standstill between New Street and Church Road through engine slipping, leading and driving sanders not working. This train, which was overloaded – 375 tons against a regulation 350 tons – was being assisted in the rear. The assisting engine for some unknown reason left the train at New Street 5

signalbox. Rule 178 carried out by fireman.' The delay was 26 minutes.

The 11.20pm Fawley-Bromford Bridge tanks was in trouble again, this time on 11 April with Barrow Road's 92007. But the crew, Barnwood Driver P E Long and Fireman J Ryder, showed initiative by dealing with the problem. The engine was short of steam due to the brickarch collapsing. The driver requested that the train be turned into the loop at Eckington, where the crew removed the firebricks from the fire, after which the driver stated 'we then carried on with no trouble at all'. The delay was 76 minutes. The report noted 'This engine had been 'out-of-course' from Barrow Road from 29.3.61 to the date of failure,' which seems to suggests that it had not visited its home depot for inspection during that period.

A similar time delay, 80 minutes, was experienced by the 10.10pm class 'C' from Nottingham to Bristol of 12 April. A Nottingham crew had Saltley's 42827, whose right hand axlebox ran hot at Bromsgrove, though it managed to get to Gloucester. The train carried on behind 44004 while 42827 was examined at Barnwood, where it was found that a lubricator pipe nut was loose, allowing oil to escape. The Saltley fitter who last examined the loco got it 'in the neck' for that casualty.

12 April was notable for the appearance of 'Britannia' 70020 Mercury on 1X05, a special from Porthcawl to Cadbury's at Bournville, and B1 61158 of 36A Doncaster on an excursion from Worksop to Gloucester.

One of Eastleigh's other 9Fs, 92206, also got into difficulties with the 11.20pm ex-Fawley, on 21 April. It was short of steam due, apparently, to having been supplied with inferior grade coal. The DMPS at Eastleigh was requested to 'ensure that the engines working this important service are completely coaled

RM/NM.110.

BRITISH RAILWAYS (WESTERN REGION)

B.R. 32852

LOCOMOTIVE CASUALTIES—NON-MECHANICAL

Station **Gloucester Barnwood.** District **Gloucester.** Date **16.5.61.**

Report of failure of Engine No. **45519** Class **6** Home Station **Bristol B.R. 82E.**
to complete scheduled working.

Driver **G. Matthews, 85C.**

Date of Failure **16.5.61.**	**Nature of Failure** **Short of steam.**
Time of Failure **2.45 onwards.**	
Locality of Failure **Washwood Hth.**	Condition of Ashpan after failure **Clean.**
Train :—	Condition of Fire after failure **Should not have caused trouble**
	Condition of Tubes after failure **Fairly clean.**
Time **2.45 a.m.**	Amount of Coal left on tender **2 ton.**
From **Washwood Heath.**	Whether fitted with S. Cleaning Smokebox
To **Bristol.**	Report on condition of tubes after failure to be completed by
Pass. or Goods. **Goods.**	boilersmith
Booked Load	
Actual Load	Foreman's Remarks **This engine is causing delay to**
State of Weather **Fine.**	**pass. and freight trains from Bristol to Derby.**
	and in my opinion requires attention
Delay 104 mins:	**urgently at shops.** **J. Barton.**

Servicing details before leaving starting points
4/10 Birmingham.

Engine worked from **15.5.61.**   Actual time engine left shed **2.15 a.m. 16.5.61.**

Actual time engine arrived on shed **21A to sgt** Date Tubeplate Cleaned **9.5.61.**

Date Tubes Cleaned *Rodded or Blown **9.5.61** Date Brickarch Cleaned **9.5.61.**

Whether fire cleaned or completely dropped **Cl.** Last Washout **9.5.61.**

Ashpan and Smokebox cleaned **Yes.**

Coal supplied	Colliery and Grade	Colliery and Grade	
Amount	**Blidworth GR.2.**		Has difficulty been experienced with this grade of coal?

* Delete as necessary.

Details of Reports Regarding Bad Steaming of Engine During Six Previous Trips

Nil at Bristol.

Conclusion **I am asking the D.T.S.O. (R.&M.) Bristol, to note this**
failure and let me have his observations on the comments contained
in the "Foreman's Remarks".

This form to be filled up in case of failure of an engine working Passenger or Freight Trains to complete its scheduled working owing to shortage of coal, shortage of steam, dirty fire, etc.   The form to be signed by the responsible Shedmaster and sent direct to Paddington and three duplicates to D.M.P.S. within 24 hours of occurrence.

Running & Maintenance Officer
PADDINGTON   Signed **For J. Powell**

---

with good quality fuel'. 92206 had been under repair at Eastleigh from 1 April to 19 April.

Holbeck's 44857 was assisting engine to 45576 *Bombay* on the 8.30am Cardiff-Newcastle express on Friday 28 April until both injectors failed at Selly Oak. It detached from the train there and ran light to Saltley mpd for examination, but somehow the repair card left with the engine was misplaced. 44857 was a busy loco, having done 102,059 miles in just over two years since an intermediate overhaul at Crewe dated 25 March 1959, a far superior yearly mileage than the two 'Jubilees' and 9F mentioned above.

The previous Friday, the assisting engine on this same working was Barnwood's 75009 before it failed at Bromsgrove with injector and steam brake problems. But 20 minutes work by a fitter at Bromsgrove shed put things right and the next day 75009 was back assisting 45602 on the same train.

4F 44102 of 82G Templecombe depot received a general overhaul at Derby Works dated 28 April and was working home on 12 May with the 6.30pm class 'C' Washwood Heath-Westerleigh and was observed by the author at Cheltenham. However it had only reached Gloucester when the left leading engine axlebox got hot. 48339 took the train on, with a 41 minute delay. The offending axlebox was reconditioned at Barnwood and the loco eventually arrived back at Templecombe on 6 June. Gloucester asked the CM&EE at Derby to 'draw this failure to the attention of the staff concerned.'

There was at least one other S&D-based loco at Gloucester on 12 May – 2-8-0 53806 was on the High Orchard branch trip working. Engines were often run in on local trippers after axlebox repairs at Barnwood, before returning to their home depots, and this was the reason for 53806's appearance on the branch. It was reported as being at Gloucester on 27 April awaiting repairs.

44102 was soon joined in the wheeldrop queue at Barnwood by 42897 of 17B Burton, whose left trailing engine axlebox ran hot, with the metal fused and underpads destoyed. It was working 4V51, the Mondays and Fridays only 1.20pm Burton-Westerleigh beer train on 15 May, giving up on reaching Gloucester, due at 4.36pm, where 44165 of Saltley took over after a 40 minute delay.

The next day Barrow Road 'Patriot' 45519 *Lady Godiva* just about managed to stagger into Gloucester on the 2.45am express goods from Washwood Heath. It was due at Eastgate at 4.26am, running time of 101 minutes, but it doubled that, losing 104 minutes on the journey, all put down to the loco. The engine had arrived at Saltley at 4.10pm the previous day, the fire, ashpan and smokebox were cleaned and it left at 2.15am as booked to run light to Washwood Heath. But it would not steam and the casualty report is reproduced. Note the Barnwood foreman's comments about this loco causing delays to passenger and freight trains from Bristol to Derby and 'in my opinion requires attention urgently at shops'. However at least in the short-term 45519 recovered as it was seen out and about on 19 and 20 May. Along with classmates 45504 *Royal Signals* and 45506 *Royal Pioneer Corps*, it was withdrawn from Barrow Road in March 1962.

On 17 May, a serious failure occurred on Saltley 4F 44333. It was working 6V04, the 4.30pm Washwood Heath-Severn Tunnel Junction class 'E' freight, seen by the author at Cheltenham before drama occurred at Lydney, where both safety plugs fused due to shortage of water in the boiler. The crew was from Gloucester Horton Road depot, not Barnwood. It is probably fair to say that Western men did not like Midland 4Fs! The blame was placed squarely on the driver and fireman – who shall remain nameless here to spare their blushes – and they were dealt with under the

Disciplinary Procedure. Meanwhile, the train, due through Lydney at 9.1pm, proceeded to Severn Tunnel Junction behind 48685 after a 250 minute delay. 44333 was taken to Barnwood, where both water gauges were found in good order, ruling out the possibility of false water level readings. At the depot, two new safety plugs were fitted, while 21 large and 15 small tubes were expanded. The casualty report stated 'Boiler examined by Derby boiler inspector on 25.5.61 who found that the repairs required were too heavy to be undertaken at the depot. The engine is to receive factory attention before being put back into traffic.'

18 May found 3F 43687 with a Bath crew having a rough trip in the early hours on a Washwood Heath-Westerleigh freight. Both Gloucester and Bristol Control complained about it, the loco losing 25 minutes over the seven mile section from Gloucester station to Standish Junction and things were no better between Charfield and Yate, where 43687 delayed the down Mail by 15 minutes. And it was carrying the wrong headlamps!

At 12.13am on 25 May at Grange Court, a week after 44333's failure, another Western driver, this time from Lydney shed, had problems with 4F 44336 through shortage of steam rather than the far more serious shortage of water. This loco, shedded at 55E Normanton and thus a long way from home, was at the head of a 6.0am (previous day) Longtown (near Carlisle) to Barry special goods, which it had worked through from Leeds. Barnwood turned out 48635 at 1.30am to take over, the delay being recorded as 155 minutes.

Canklow depot's 73043 suffered on Saturday 27 May with a broken blower pipe in the smoke box while working the 10.45am class 'A' passenger from Sheffield to Bournemouth, so it retired at Gloucester in favour of classmate 73155. The blower pipe was

Below: As mentioned in the narrative, the Cardiff-Newcastle was double-headed from Gloucester on Monday 29 March 1961 by 92139 and 73043 and is seen at Bromsgrove before going up the Lickey incline. 73043 was having blower problems and came off the train at Birmingham New Street.
R K Blencowe collection

brazed and refitted at Barnwood. The loco remained there all Sunday and worked the morning Cardiff-Newcastle express on Monday. As the train was just over the regulation load for one engine of 350 tons, assistance was provided by Saltley 9F 92139, which had also spent Sunday at Barnwood. The 9F probably did most of the work on the journey because Driver W Dipple reported 73043 was still having blower problems and it went to Saltley depot for attention after giving up at Birmingham New Street. 73043 had covered 106,200 miles since its last shopping, an intermediate overhaul at Doncaster dated 5 July 1958.

Going back to 27 May, Millhouses 45576 *Bombay* had charge of the 4.45pm mail from Bradford to Bristol, until it became a casualty, Driver Leadbetter of Bristol reporting the engine 'riding very rough'. 75022 took the working on from Eastgate, with a delay of 33 minutes. 45576 was examined at Barnwood, where it was seen on the Sunday, but no defect was found, and it went back home light engine with a repair card.

27 May was noteworthy for the appearance of former Barnwood allocated preserved Compound 1000 which came down light in the morning to work a railtour north from Gloucester. Also of interest was 5D Stoke shed's 42392, down light engine; rebuilt 'Patriot' 45512 Bunsen of 12B Carlisle Upperby on a pigeon special; and IA Willesden

engine, Fowler-tendered 8F 48603, on a down freight, so quite an interesting day at the lineside.

On Tuesday 6 June, the 11.45pm Bristol to Derby parcels, reporting number P486, was well loaded, as usual, with 19 vehicles totalling 471 tons. In charge was Derby's 'Jubilee' 45557 *New Brunswick*, unfortunately not in the best condition. Driver F E Ford and Fireman J Banks reported steaming problems with a collapsed brickarch and defective exhaust injector. What with these problems and various permanent way slacks, departure from Gloucester 23 minutes late became a 60 minute late arrival at New Street, at 3.28am on 7 June. 45557 had been on report at Bath on 13 May for leaking tubes and again on 16 May with a broken tender spring, replaced there by one off 45558 *Manitoba* – a rare visitor – which had itself failed with injector problems. 45557 went back to Barnwood later on 7 June, where a new arch was fitted; large boiler tubes expanded; a burst superheater element replaced; and small tubes cleaned. This parcels was always a heavy train which could be heard approaching from miles away – a Gloucester fireman on a freight working stopped by signals at Westerleigh in the early hours recalls hearing this train roaring out of Bristol with a 'Jubilee' in full cry, always a fine sound!

73043 seems to have recovered from its defective blower of

DRIVER'S/MOTORMAN'S REPORT

BRITISH RAILWAYS   Glos.85C   Depot   W.   Region. Date   June 13th   19 61   B.R. 32841

Date of Incident ... June 12 19 61   Place of Incident ... Stoke Wks Jct.

Driver/Motorman }* L.A.Lord   Reg. No. 120   Depot Glos.85C

Fireman/Second Man }* C.Phillips   Reg. No. 253   Depot " "

Guard H.Shurmer   Station Glos.85C

11.50 a.m. Pssr. }* Frt. Train from Barnwood to W.W.Heath

on June 12th 19 61   Reg. load 99 Gds   Actual load 73 Gds (approx)

Locomotive/Diesel/Power Car Electric Multiple/Diesel Elec./Unit. }* No.   Train 48063   Assisting

Class/Type* 8 Frt

*Delete words not applicable

I worked the above train from Barnwood to Bromsgrove. This was a very heavy train consisting of 29 16 tons wagons well loaded with slack coal next to the engine, and 24 grampus wagons empty behind. I was brought to a stand at Stoke Wks outer home signal which was in the on position, at approx 1/25. The line from Stoke Wks outer home signal to the signalbox is a curving incline, and when the signal was pulled off at approx 1/35, owing to the steepness of the road and the weight off the train I could not pull the train forward. I reversed two or three times to close the wagons against the Guards brake to try and start the train, but the Guards brake wouldnt hold the train and I only succeeded in pushing the brake further back. I then phoned the signalman at Stoke Wks signalbox and explained the position to him and asked him to arrange for an engine from Bromsgrove to assist me. The assisting eng. arrived at 2/0pm and the train left at 2/3 arriving at Bromsgrove at 2/12 where I was relieved by Saltley Dr.Lloyd. I was informed today by this driver, that with equivalent to 3 bankers it took him 40 minutes to travel from Bromsgrove to Blackwell for which distance this train is which distance allowed 15 mins.

(Signed) L.A.Lord.

late May and headed the Sundays M318, 8.15pm Gloucester-Catterick Camp class 'A' passenger on 11 June. But it encountered an unusual fault en route. Barnwood Driver A John wrote 'When passing Kingsbury, Fireman M Limbrick could not open firehole door, therefore stopped at Burton to have them put right.' The difficulty was caused by a bolt lodged between the firehole door and the baffle plate, so removal of this solved the problem.

8F 48063 was in trouble on 12 June with the 11.50am Barnwood-Washwood Heath freight. Driver L A Lord's report is reproduced. In fact, thanks to a timetabled layover of nearly an hour at Eckington for other trains to pass, the freight actually arrived at Bromsgrove 11 minutes early, despite the problems encountered. But it held up traffic on the Lickey incline, taking so long for the ascent.

This was right at the start of the summer timetable and the first day on which shiny new Sulzer Type 4 diesel locos, eager to show their paces, especially on the Lickey's gradients, were diagrammed for express passenger trains on the line. D102 had the 7.40am Bristol-Bradford that morning, while other initial diesel turns were the 12.48pm York-Bristol, 7.5pm Newcastle-Bristol and 8.15pm Bristol-Derby parcels.

'WD' 8F 90573 was having steaming problems on 15 June. Barnwood Driver C Reed and Fireman Mike Randall had a rough trip with it on the 2.55am Cardiff-Woodford Halse as far as Honeybourne, the driver reporting 'When at Honeybourne informed Control that fire was in dirty condition and wanted cleaning. After doing all we could to the fire with fire irons, informed Control we would do the best we could. No time lost between Honeybourne and Gloucester South on the return.' The much delayed return was on a Woodford Halse-Stoke Gifford freight, which upon arrival at Gloucester South at 4.5pm was 268 minutes late. The crew took 90573 to Barnwood for fire cleaning and then the loco proceeded on its train with a fresh crew. Mike Randall recalls that the coal in the tender was very poor quality.

9F 92221 gave up short of steam at Bromsgrove on 21 June on the previous night's 11.30pm Avonmouth-Bromford Bridge tanks. Bromsgrove banker 92079 was then utilised, still with the Barnwood crew of Driver G Matthews and Fireman D Norman, to take the train to its destination. Arrival at Bromford Bridge was 6.40am, 96 minutes late. (It would be interesting to know if Bromsgrove's famous 'Big Bertha' 0-10-0 58100 was ever called upon in similar circumstances, taking it away from from its normal banking duties.) Driver Matthews spent time as Mayor of Gloucester – one of quite a large number of loco drivers who held civic office throughout the UK.

Another 9F gave up a day later while hauling 9V30, the 10.10pm Washwood Heath-Stoke Gifford. The loco, 92130, was from 18A Toton depot and ran out of coal at Stonehouse. Perhaps it was a late replacement for another engine and had not been coaled up by Saltley depot when called out for this duty. It left the train at Stonehouse and ran light to Barrow Road shed.

Dursley Driver E Carpenter earned praise for his actions on 23 June. With Fireman Humphries, he was running engine and brake van on the short branch from Dursley to Coaley Junction with regular loco 0-6-0PT 1605, when the left-hand rear portion of the side rod became disconnected at the gradient joint. As he said 'I took the other side off and proceeded to Coaley Junction. Time (of incident) 8.15am, arrived Coaley Junction 9.53am.' 1605 had already done a mixed and a freight trip on the line, but due to the problem, the 8.25am Coaley-Dursley passenger and the 9.55am return were cancelled, this being a one-engine-in-steam branch. 6437 was sent from Gloucester Horton Road to take over.

A fitter's mate was sent to walk the branch from Coaley to Dursley and back to find the parts which had gone missing, but he only found the gradient pin close to the over bridge near Workman's Mill. The full report and Mr. Powell's commendation are reproduced

There is a tale related by Fred Cole and written up in the Oakwood Press book on the Dursley branch that, in February 1961, an inexperienced fireman allowed the 1600 class loco to run out of water at Coaley. The story made the national press because the six waiting passengers formed a chain to pass buckets full of water to the engine's tanks! These shenanigans not surprisingly earned the engine crew a reprimand from management!

A 'foreign' engine arriving at Gloucester in the early hours of

Right: 0-6-0PT 1605 was a regular performer on the Coaley Junction to Dursley branch. On 20 June 1961 it propels its mixed consist towards Draycott flour mill seen in the background and connected to the branch by a short curved siding. This morning train was shown in the working time table as a freight and the photographer, Ben Ashworth, confirms the branch coach was indeed empty stock. Ben also recalls the driver, Ernie Carpenter, mentioned the loco's tubes were 'spitting' but it was another problem which failed the loco just three days later. The driver's initiative on that occasion was rewarded with a commendation from the District Traffic Superintendent. *B J Ashworth*

BRITISH RAILWAYS

B.R. 87315

LOCOMOTIVE CASUALTY REPORT (MECHANICAL)

Locomotive Number	1605

Motive Power
Running & Maintenance Depot **Dursley** District **Gloucester** Date Initiated **27.6.61.**

Locomotive No. **1605** Class **2F** Allocated to **85C** Date of Casualty **23.6.61.**

Driver **E. Carpenter** (No.) **21** Fireman **Humphries** (No.) **26** Stationed at **Dursley**

Working the **8.10** a.m Class **'G' E&B** Train from **Dursley** to **Coaley**

on **Fri** day, the **23rd** day of **June** 19 **61**.

Assisting { Locomotive No. .... Class .... Allocated to ....

Assisted by { Driver .... (No.) .... Fireman .... (No.) .... Stationed at ....

became a casualty at **Cam** causing a delay of **98** mins. Locomotive changed at **Coaley**

No. and Class of Locomotive working forward **6437** Load of train **E. & B.** Regulation load
for locomotive

NATURE OF CASUALTY

Left coupling rod disconnected at gradient joint

CAUSE OF CASUALTY (Full description)

Unable to state cause of "Joint Pin" working out of position, until
Safety Bolt and nut, also flush fitting washer has been located. Further
search of line between Coaley Junction and Dursley to be carried out.

PARTICULARS OF REPAIRS CARRIED OUT NECESSITATED BY CASUALTY

R. & L. H. Coupling Rods and Joint Pins sent to Swindon for repars.

HISTORY OF LOCOMOTIVE NO. 1605

Date and classification of last Shop Repair when the part } **24.4.56.** at **H/G Swindon** Works
affected received attention

Estimated mileage since **75,250.**

*Date of last "X" Examination **19.5.61.** at **Gloucester** Depot

Working days since ....

*Date of last Washout .... Working days since ....

*Date of last Daily/Weekly Examination **19.5.61.** at **Gloucester** Depot

by **A. Lowe** (No.) **463** Grade **Exam. Fitter**

*Date of last Periodical or Mileage Examination **4.7.60.** Item No. **14** at **Gloucester** Depot

Estimated period/mileage since .... Extent overdue ....

Date defective part last } **19.5.61.** at **Gloucester** by **A. Lowe** (No.) **462** Grade **Ex Fitter**
examined or renewed

*If relevant.

YOUR REF.	—	BRITISH RAILWAYS	OUR REF.	RM/M.216.
DATED			DATE	25.7.61.

BRITISH TRANSPORT COMMISSION
BRITISH RAILWAYS

26 JUL 1961
W.R.
GLOUCESTER (BARNWOOD)

B.R. 14302/252

DISTRICT TRAFFIC SUPERINTENDENT
WESTERN REGION
GLOUCESTER

TO Mr. F.A. Cole,
Gloucester Barnwood

Telephone: GLOUCESTER 21121
Extn. 74.

Engine 1605 (85C) Mechanical Casualty at Cam,
23.6.61. 8.10 a.m. 'G' Dursley to Coaley.

With reference to the above and my concluded casualty
report of the 19th instant, I shall be pleased if you will
arrange to see Driver E. Carpenter of Dursley, and convey to
him my appreciation of the initiative which he displayed in
connection with this incident when he removed the other portion
of the Left Hand Side Rod which enabled the engine to proceed
to Coaley Junction without assistance, thus saving the Break Down
Van being sent to the scene.

FOR J. POWELL

District Running and Maintenance Officer

Date	Loco	Home shed	Train	Problem	Place	Delay (mins)	Notes
10/04/61	45598	17A	7.25pm Bristol-Newcastle class A	INJ	ER	3	
13/04/61	45506	82E	9.2pm Bradford-Bristol class A	INJ	ER	5	
16/04/61	45654	41C	7.5pm Newcastle-Bristol class A	SAND	BNS	7	
16/04/61	45626	17A	7.45pm Bristol-Catterick Camp class A	SOS	ER	10	1
19/04/61	45569	55A	10.45am Manchester-Bournemouth class A	SOS	ER	7	
21/04/61	44171	21A	1.45am WWH-Westerleigh class H	MECH	GLOS	35	2
24/04/61	44659	21A	1.10am Bristol-Sheffield class A	HOT	CHEL	7	3
25/04/61	45618	17A	9.45am Bournemouth-Manchester class A	SOS	GLOS	28	4
26/04/61	45260	21A	12.48pm York-Bristol class A	SOS	GLOS	4	5
28/04/61	45651	82E	9.30am Paignton-Bradford class A	MECH	GE	7	6
30/04/61	45627	41C	8.0am Bristol-York class A	PRI	ER	23	7
02/05/61	42756	17B	7.0pm Bristol-Derby freight	MECH	GLOS	20	8
12/05/61	92029	15B	9.20am Edinburgh-Stoke Gifford goods	COAL	GLOS	35	9
16/05/61	45612	17A	12.43pm Newcastle-Bristol class A	SOS	ER	10	10
28/05/61	44185	21A	8.15pm WWH-Westerleigh freight	SOS	ER	144	
31/05/61	90573	88A	2.55am Cardiff-Woodford Halse goods	SOS	ER	45	
05/06/61	45610	17A	6.18am Derby-Bristol class A	INJ	ER		11
06/06/61	44851	17A	12.43pm Newcastle-Bristol class A	HOT	ER		12
08/06/61	43507	21A	3.15pm WWH-Gloucester class J	HOT	CHEL	15	13
13/06/61	92052	21A	11.20pm Fawley-Bromford Bridge class D	COAL	GLOS	7	
14/06/61	44436	17B	4.20pm Burton-Bristol class D	HOT	GLOS	8	14
15/06/61	45590	41C	9.2pm Bradford-Bristol class A	MECH		15	15
23/06/61	43949	21A	7.43pm Bristol-Bescot fitted freight	SOS	GLOS		
25/06/61	44858	21A	9.45am Bournemouth-Manchester class A	INJ	GLOS		16
26/06/61	44004	21A	8.55pm Avonmouth-Water Orton class D	SOS	ER		17
29/06/61	45557	17A	10.25am Manchester-Bournemouth class A	SOS	ER	5	
29/06/61	92014	18A	2.30pm Barnwood-Lawley Street fitted freight	MECH			18

Notes

1	Engine changed at Gloucester
2	Big end knocking badly, repaired at Barnwood. 44264 took train onwards
3	44810 worked train north. 44659 light to Barnwood for attention
4	Fresh loco from Gloucester. 45618 examined at Barnwood
5	45607 worked train to Bristol. 45260 sent to Saltley for repairs
6	Axlebox lubricator failed. 45657 worked train forward
7	44920 worked train northwards
8	Steam brake repaired at Barnwood, loco worked on
9	Only 2 cwt of coal left in tender on examination at Barnwood
10	Brickarch collapsed, engine home light. 75022 worked train
11	Loco repaired at Barnwood. 44873 worked train
12	Loco repaired at Barnwood. 45006 took train
13	Loco light to Barnwood for repair
14	75009 took train on. 44436 repaired at Tyseley
15	Blowdown pipe fractured. 44963 worked from New Street
16	45264 worked forward. 44858 repaired at Barnwood
17	44004 light to Saltley for attention
18	Loco failed at Defford. No further details

Saturday 24 June with the 9.5pm Newcastle-Paignton passenger was 'Black Five' 45097 of 24L Carnforth, well away from its regular haunts, which was not at all uncommon on summer Saturdays. It worked the train from Sheffield, but got no further than Gloucester due to a hot axlebox on the tender, so spent a while at Barnwood under repair – it was there on 16 July, but had gone by 30 July. The replacement loco would also have been rare a couple of years earlier, being 'Royal Scot' 46118 *Royal Welch Fusilier*, but now a regular sight on the route, having transferred to 16A Nottingham in December 1959 from 1B Camden, then on to 17A Derby. By June several of the class were also allocated to 21A Saltley.

46118 later worked north on 1N84 2.15pm Bristol-York, while another 'foreign' 'Black Five' seen that day was 45466 from 12A Carlisle Kingmoor on IM07 Saturdays only 10.32am Bournemouth-Manchester Victoria.

JULY – SEPTEMBER 1961

'In view of the serious delay...I am arranging for my Gloucester Barnwood shed master to follow up ... with the Barnwood Preparation Driver.'

The first half of July saw a number of Saltley locos feature in the casualty reports. On 1 July, 48669 had defective injectors and was declared a failure by Barnwood Driver P F Preece at Cheltenham High Street on the 4.28pm Gloucester-Lawley Street class 'J' goods. Shedmate 48315 took over after a 50 minute delay.

44463 failed on the 1.30pm Washwood Heath-Bristol freight at Stonehouse at 8.0pm on 3 July. The cause was dirty tubes and inferior quality coal. 44463 had enjoyed a fairly leisurely time recently according to its service details received from Saltley. Arriving on shed there at 6.30pm on 19 June, it did not leave again until 1.0pm on 3 July, for this working; during its sojourn the tubes and tubeplate were recorded as having been cleaned.

On 5 July, the passengers on a special from Bristol worked by 44818 would no doubt have been getting restless after the loco failed at Gloucester with a defective exhaust injector. The delay is recorded as 108 minutes, before the replacement engine, 92156, with Barnwood Driver J Lees and Fireman M Davis, took the train on its way. The special was going to Thurso in the north of Scotland so had an extremely long journey ahead of it, without having to contend with engine failures as well. It is shown in the report as a class 'B' train – hopefully not stopping at all stations on the way to Thurso! It would be interesting to know what type of special this was – possibly military.

On 11 July, 44091 lost 1 hour 20 minutes between Barnwood and Bromsgrove on the 7.0pm Bristol-Derby goods. Not all the loss was due to the loco according to 85C's Driver Matthews who was with Saltley Fireman Dugard. 33 minutes was spent at a signal check at Dunhampstead. But the loco's fire was very dirty and the injectors were weak, so 44165 took over at Bromsgrove with a crew from that shed.

The next report, dated 13 July, was prepared by Mr A Core, the Bromsgrove depot shed master. Saltley's 44013 had the 11.50am class 'H' from Barnwood to Washwood Heath with

Driver Grant. At Spetchley he realised the right big end cotter was missing, so that brought proceedings to a halt, and it was exactly 1223 minutes – over 20 hours – before the freight moved off hauled by 92157. As for 44013, a spare cotter was obtained from Saltley, fitted, and the engine went home light for further attention. Mr Bartlett, under whose command Bromsgrove came, it being in the Gloucester district, wrote 'In view of the serious delay ... I am arranging for my Gloucester Barnwood shed master to follow up ... with the Barnwood Preparation Driver.' This was Driver Bill Dipple, a top link man when mentioned in 1960 reports. He stated that both cotters had been present when he prepared the loco.

Not many days passed before 92157, just mentioned above, needed rescuing itself. It had the 10.10pm Bromford Bridge-Fawley oil empties on 18 July, but failed at Gloucester in the early hours of the next day with defective injectors. Classmate 92004 took over, with a delay of 82 minutes recorded.

On 27 July, full tanks on the 11.20pm ex Fawley were held up for 37 minutes when Eastleigh's 92231 failed with a hot axlebox. This heavy train would have made a stirring sight heading north from Gloucester behind 92231's replacements, doubleheaded 4Fs Nos. 43975 and 44226.

The same train suffered a 15 minute delay on 2 August when 92204 failed short of steam due to a very dirty fire and heavy clinker. After attention at Barnwood, it went light to Saltley for the return working.

12 August witnessed the last passenger trains on the Ashchurch-Upton-on-Severn branch, 43754 being on the final working. Although a freight service continued, it sounded the death knell for the small shed at Tewkesbury. The engine for the branch freight, and the loco off the Evesham line passenger which had stabled overnight at Tewkesbury, used Cheltenham Malvern Road depot instead.

'WD' replaced 'WD' on 16 August on the 6.35am Woodford Halse-Stoke Gifford freight. Barnwood men took over 90504 from a Woodford crew at Honeybourne, but on leaving, discovered the anti-vacuum valve was fractured, later found to have been caused by a faulty casting. On arrival at Gloucester, they took the loco to Horton Road shed, which provided 90095 as replacement, the train being held up for 63 minutes. The defective valve was sent to Crewe Works for repair.

Barnwood's own 44296 had steaming problems on 18 August with a Washwood Heath-Westerleigh working, after which the Gloucester DMPS blamed Saltley for lack of servicing.

A highly unusual happening on 20 August was 45519 Lady Godiva *turning up at Eastleigh with the Fawley empties, no doubt covering for a failed 9F.*

Eastleigh's 92239 was on the 2.20pm Fawley-Bromford Bridge tanks on 24 August, but had to come on shed at Barnwood for more coal, causing the Gloucester DMPS to take up that particular problem yet again with his counterpart at Eastleigh.

On 25 August he was complaining about unsatisfactory servicing by the Eastern Region's Sheffield Millhouses depot. 45557 *New Brunswick* had the down 'Pines Express' to Gloucester when it failed short of steam with a dirty fire, full smokebox and full ashpan, indicating lack of attention during its five and a half hour layover at Millhouses earlier that day. 45557 returned north the next morning with 1N27, the 8.40am Bristol-Newcastle.

That day, 26 August, the down 'Pines Express' was seen at Hatherley behind 9F 92000, passing Eastleigh-based 92206 going north on a relief passenger ex-Bournemouth, which it worked through to Manchester – showing the reliance being placed on 9Fs to do main line passenger work. Even so, 92000, based at Bath shed, was booked for repairs there later that day with badly leaking tubes.

A Saltley crew managed to get into serious trouble with the recently mentioned 44296 on Sunday 27 August. They had the 12.45am Westerleigh-Washwood Heath class 'F" goods which was travelling via the Ashchurch-Evesham-Barnt Green route, possibly due to weekend engineering work on the Bromsgrove main line. At Redditch shortage of water in the boiler meant that the lead plugs fused. As a result, the chief boiler inspector examined the loco and several roof nuts in the firebox had to be changed.

Engines from 18A Toton figured prominently in the next few reports. On 31 August 92052 worked from its home on the 12.36pm goods to Westerleigh, with Gloucester Driver G R John from Bromsgrove. It struggled with a dirty fire and shortage of coal. Examination by Foreman R W Davis at Barnwood found 'Firebox full of rubbish. All slack coal, very poor, 2 cwt left on tender.' 48186 – another Toton loco – worked forward, with the delay due to 92052 shown as 2 hours 42 minutes.

Toton engines were still in the news on 5 September, when 92056 stopped outside Barnwood for a washout plug to be tightened while in charge of the 11.0am Avonmouth-Bromford Bridge tanks, with just a minor delay. Later that day, shedmate 92055 took over at Gloucester from another Midland line 9F, 92154 of 15A Wellingborough. The latter had the 3.25pm Washwood Heath-Westerleigh, but failed with leaking tubes at the firebox end. The delay was 248 minutes.

Moving on to 7 September, Toton's 92078 found itself elevated to express passenger work. Barrow Road's 'Jubilee' 45662 *Kempenfelt* arrived at Eastgate on the 2.15pm Bristol-York class 'A', with the middle cylinder steam pipe joint blown out. While 92078 headed north on the express, 45662 returned light to Bristol for attention.

Cheltenham Malvern Road depot played host to 88A Cardiff Canton's failed 'WD' 90572 on 8 September. It was working the 2.55am Cardiff-Woodford Halse, load 42, with Barnwood crew H J Clarke and W McGhee. The driver reported 'Lubricator condensing pipe split, also lubricator delivery pipe joints blowing in engine cab. Fitter on Malvern Road unable to do repairs, so engine left on loco.' A fresh engine, 3838, came over from Gloucester to take the train forward, the booked delay being 210 minutes.

A day later, Saturday 9 September, Malvern Road shed played host for the last time to a Southern Region engine off the Southampton-Cheltenham passenger working over the MSWJ route. The loco was class 'U' 2-6-0 31791 of 71A Eastleigh.

90572 was subject of another casualty report when it failed on 20 September. Working the same train as previously, it got to Honeybourne this time, where the Barnwood men, Driver

C Reed and Fireman Cotterell reported 'When arriving at Honeybourne found pin on left engine leading spring hangar one and a half inches out of place. Informed Control. Took engine to Stratford loco shed, at reduced speed, for repairs.' At Stratford, Fitter H Duckett refitted the pin and replaced a missing split pin, which was the cause of the defect. The engine then went light to Tyseley depot, Birmingham, as required by Engine Control.

The 2.20pm Fawley-Bromford Bridge tanks continued to have bad days. On 18 September, Barrow Road's 92007 had to go to Barnwood shed for servicing, because of a dirty and clinkered fire. Once again, Mr Bartlett complained to his opposite number at Eastleigh.

A notable sighting on the 2.20pm ex-Fawley on 21 September was 'Jubilee' 45662 Kempenfelt, officially transferred from Bristol to Shrewsbury this month.

On 28 September, Saltley's 92136 was on the same turn. When Barnwood Driver H Burrows and Fireman J R Cooke took over at Dr. Day's Sidings, Bristol, having worked down on the 3.0pm ex-Bromford Bridge, they were informed the engine was not fit to work forward and that Control had been asked to provide another loco. Driver Burrows takes up the tale: 'Part of the brick arch was in the fire, which was very dirty and the tank almost empty. I went to the signalbox and informed the signalman I should require water at Lawrence Hill and assistance from Stoke Gifford. He told me to wait until he could get a path for me. Assistance was given from Stoke Gifford to Barnwood, where I left the engine on the loco and reported the fire and brick arch.' Delays were shown as 69 minutes at Dr. Day's (though 39 minutes was due to the crew being late off their inward working); 10 minutes water stop at Lawrence Hill; and 32 minutes for steam at Stoke Gifford.

OCTOBER – DECEMBER 1961

'...a lot of water coming out through the ashpan, also firebox leaking both sides – large and small tubes leaking.'

Locos sustaining hot boxes at various locations found their way to Barnwood for attention. One such was 'Castle' 5049 *Earl of Plymouth* which had the 11.40pm class 'A' passenger from Paddington to Bristol until it got a hot box at Swindon in the early hours of 1 October. The loco went to Barnwood, with the wheels and boxes being returned to Swindon Works for fettling. 5049 was ready to resume duty on 7 November.

Two locos succumbed at Gloucester in the early hours of 18 October. Saltley allocated 'Jubilee' 45579 *Punjab* was on the 10.19pm Nottingham-Bristol St. Philip's class 'C', 4V35, but was failed with 'alleged defective injectors' though examination at Barnwood could find nothing wrong. 73093, probably Eastgate station pilot, took over the train. 'Royal Scot' 46103 *Royal Scots Fusilier*, also a Saltley engine, had the 7.5pm Newcastle-Bristol mails, but was short of steam and also could not heat the train. So it was replaced by 73136. 46103 had covered 36,000 miles since its last shop repair at Crewe dated 24 January 1961. Both the 'Jubilee' and 'Royal Scot' types were recent transfers to Saltley, which did not have these classes on its books at the start of the 1960s. This showed once more the increasing effects of

OTHER CASUALTY REPORTS July-September 1961

Date	Loco	Home shed	Train	Problem	Place	Delay (mins)	Notes
15/07/61	44849	55A	10.5pm Hull-Paignton class A	PRI	ER.	8	
18/08/61	44296	85C	11.18am WWH-Westerleigh freight	SOS	ER.	36	
21/08/61	73003	82E	4.45pm Bradford-Bristol class A	INJ	ER.		1
22/08/61	42827	21A	9.45am Ashchurch-Birmingham class J	HOT		45	2
27/08/61	45668	17A	12.15pm York-Bristol class A	SOS	ER.	20	3
30/08/61	90524		2.15am Rogerstone-Woodford Halse freight	INJ			4
05/09/61	48336	21A	11.20pm Stoke Gifford-Bordesley freight	MECH	CHEL	90	5
06/09/61	44263	21A	8.55pm Avonmouth-Water Orton freight	MECH	GLOS	46	6

Notes

1	73003 repaired at Barnwood. 46148 took train on to destination
2	43887 worked train from Ashchurch. 42827 light to Saltley for attention
3	45668 off train at Duddeston Road to Saltley mpd. 45602 took over
4	Loco failed at Bishops Cleeve
5	Left hand piston gland packing blown, to Barnwood for repair. 90504 worked train
6	Blast pipe joint blowing in smokebox, loco light to Saltley for attention. 92152 took over the train

DRIVER'S/MOTORMAN'S REPORT

BRITISH RAILWAYS

B.R. 32841

Gloucester .... Depot .... W. .... Region. Date ...... Fri. 20.10.61 ....19......

Date of Incident ...... 20.10.61 ......19...... Place of Incident ....

Driver/Motorman }* W. Mellor ...... Reg. No. ...... 16 ...... Depot ...... Glos. B'Wood.

Fireman/Second Man }* L. Evans ...... Reg. No. ...... 280 ...... Depot ...... " ...... "

Guard ...... ? ...... Station ...... New Street

2/50 a.m./p.m. Passr./Frt. }* Train from ...... Worcester ...... to ...... New Street.

on ...... Fri. 20.10.61 ......19...... Reg. load ...6... Vehicles ...... Actual load ...180... Tons.

Locomotive/Diesel/Power Car Electric Multiple/Diesel Elec./D.E. Unit. }* No. ...... 45221 / 21A.

Assisting ....

Class/Type* ...... 5P.

*Delete words not applicable

Time Lost by Loco.

I have to report that after taking to Engine 45221 at Worcester we proceeded to Droitwich where there were signs that we should be in trouble, we arrived halfway into Bromsgrove Station where we came to a Halt 10 inches Vacuum ,100 lbs Steam and water very low, the fire was solid and the only fuel we had available was welsh slack.We had no rake ,obtained one from Bromsgrove and the Dart was of little value, the Bromsgrove Foreman would say there was a large fire in the firebox which was solid, my fireman put very little coal on as it was putting coal onto coal and would have been useless we losr about 15 minutes. for your information. Will.Mellor.

Signature ....

Above: One of the Midland types which survived pretty well at Barnwood over the years was the 'Three Freight' or 3F 0-6-0. 43645 was a long-time Barnwood resident and received a heavy intermediate overhaul at Derby Works in September 1961, though that did not prevent it suffering a mechanical defect a few weeks later between Ashchurch and Evesham. In this picture at Hatherley it is scurrying home to Gloucester, probably from duty at Ashchurch. Note the surplus track and fittings which had come from the recent removal of Hatherley Junction.   *R Stanton*

dieselisation – 46103 had previously been at work on the Midland main line out of London St. Pancras, allocated then to 14B Kentish Town, while 45579 was formerly at 17A Derby and was a regular on the Birmingham-Bristol route.

BR Standard 4-6-0s 73091, 73092 and 73093 were recent transfers to Barnwood from Shrewsbury – where they were displaced by former Bristol Barrow Road 'Jubilees' made redundant by dieselisation. These class 5 locos regularly did Eastgate station pilot/standby turns in place of the class 4s previously utilised – as illustrated in the casualty reports the pilot was often called upon for express passenger duties. While 73092 and 73093 were both noted on Barnwood in mid-September, 73091 appears to have come into service later, as it spent several weeks in October and November 1961 stored at Cheltenham Malvern Road shed. Barnwood's three displaced class 4s – 75002, 75009 and 75023 – moved to Templecombe, the latter two on 9 September, for Somerset & Dorset line duties where they once again took over from Midland 4-4-0s, having done that when they were originally transferred to Gloucester. And after a stint on the S&D, all three transferred to the Cambrian Coast lines.

There was some confusion later on 18 October at Worcester. Barnwood Driver Wilf Mellor and Fireman L Evans worked in with 45668 *Madden* on the 1.45pm passenger from Gloucester. As rostered, they changed engines with a Worcester crew, taking over 'Crab' 42823 to work the passenger forward to Birmingham. But 42823 could not maintain the vacuum. The driver requested fitter's attention, but the station inspector had already asked for another engine. The upshot was that the Barnwood men ended up with their original charge, 45668, and the train went away 20 minutes late.

Two days later, the same crew again had problems on the working. Having taken 45610 *Ghana* to Worcester, they received 45221 but this only got to Bromsgrove. The driver's report is copied.

On 19 October, 92248 was on the 11.20pm Avonmouth-Bromford Bridge tanks, but got no further than Gloucester after a tube burst. 92110 took over the working. Just a few days later, on 23 October, 92248 became a total failure with the 11.0am Avonmouth-Bromford Bridge tanks. At Eckington the crew noticed another burst tube and within 40 minutes the fire was completely out. The train went on after a 72 minutes delay behind 8F 48220 which had been on its way light engine to Saltley. Meanwhile 92248 was towed dead to Barnwood.

On 27 October Barnwood, and former Tewkesbury, Driver Paddy Turberville and Fireman R Shaw worked the 3.25pm Ashchurch-Evesham freight, load 17 wagons, with 3F 43645. But they had only travelled the three and a half miles to Beckford before a problem arose. The driver could not shift the reversing lever. It appeared the left die block had seized and the driver needed the help of his Ashchurch guard F Sweet to get the lever into forward gear. But then they could not get it into reverse, so failed the engine. Classmate 43593 was summoned with another Barnwood driver and Fireman Mike Randall and the two engines took the train to Evesham. What a fine sight that made! At Evesham 43645 was put on the turntable and eventually the two locos went back to Barnwood. 43645 had only just received a heavy intermediate repair at Derby Works dated 21 September, since when its estimated mileage was just 509. It was repaired at Barnwood, returning to service on 13 November.

From 6 November 1961, Sulzer Type 4 diesels were diagrammed for some of the fast freights on the Birmingham-Bristol line. But one of these diesel workings, the class 'C' 2.55pm Bristol-York Dringhouses, reporting number 4N57, was steam on 7 November, hauled by Saltley's 7P 46137 *The Prince of Wales's*

51

Above: 'Royal Scot' 4-6-0s became an everyday sight for a while in the latter years of steam between Birmingham and Bristol. In this photograph 17A Derby's 46137 looks in good order on a down express. It had been overhauled at Crewe Works in January 1961. By November 1961 it appears from a casualty report to have gravitated to Saltley and managed to become a failure on one of the most important express goods trains on the line, the 2.55 pm Bristol-York, on 7 November. This train had been one of the first fast freights in the area to be scheduled for Sulzer type 4 diesel haulage – ironically from the previous day, 6 November.

R Stanton

Volunteers (South Lancashire). However the Barnwood driver complained that heavy blows of steam at the front end prevented him seeing the signals, so he called for a fresh engine from Gloucester, which was 73094. Like shedmate 46103, 46137 had a shop overhaul at Crewe in January 1961, presumably prior to its transfer to Saltley, and had since worked 25,500 miles. 46137 had been the last unrebuilt 'Royal Scot', not being attended to until March 1955.

The 2.20pm ex-Fawley was in trouble again on 8 November. At 11.45pm Barnwood shed received a message from Control asking for an engine to go to Cheltenham High Street and assist the train. The depot turned out 44813 just after midnight which assisted 92216 to Birmingham Landor Street, from where 92216 went unaided to Bromford Bridge.

On one occasion during November, 'Royal Scot' 46162 Queen's Westminster Rifleman *was seen on the Fawley empties.*

Millhouses 'Jubilee' 45664 *Nelson* seems to have disappeared for a while in November. It worked the Sundays 9.25am Derby-Bristol express on 12 November, but gave up at Gloucester. It was using excessive amounts of water, with Driver G Hewett reporting 'a lot of water coming out through the ashpan, also firebox leaking both sides – large and small tubes leaking.' 44810 took the train to Bristol, while 45664 left Barnwood shed at

3.0pm light engine for its home depot. Barnwood Shed Master Fred Cole wrote: 'Passed Fireman N Sutton worked the engine to Saltley taking a repair card with him, which, he assures me, was handed in at the Foreman's office, Saltley. He also acquainted the Assistant Running Foreman that the engine was en route to Millhouses for repair.' But over a week later, on 20 November, the Millhouses shed master stated that 45664 had not yet arrived back. Where was it? Perhaps Saltley borrowed it for their turns, not at all unknown! On the other hand, the loco might have got so bad that Saltley stopped it there for repair and were just a bit dilatory in doing the paperwork.

Barrow Road's 73031 had an easy job on 16 November, the 4.20pm slow passenger from Bristol to Gloucester. Nevertheless the train had to be terminated at Berkeley Road when both injectors failed. The stranded passengers were presumably transferred to the 5.15pm Bristol-Birmingham stopper. The disabled engine and coaches were towed to Gloucester by 73093.

The 11.20pm ex-Fawley tanks was still recording delays. On 14 November, Saltley's 92139 was replaced at Gloucester by 92136. 92139 had the usual problems of a dirty fire and part of the brickarch missing. It received a new arch, made of concrete, as was usually the case with Saltley's 9Fs.

Toton's 92156 was on the same job on 20 November, its problems were not just shortage of steam, but also defective injectors. A water test revealed six leaking elements, while 30 small and 28 large tubes were blocked up. 48060 took the tanks to Bromford Bridge and 92156 went home light engine with a repair card for attention to the elements.

On 22 November, Saltley 'Black Five' 44841 ran light from Barnwood for the 2.0pm class 'C' Evesham to Water Orton, load 39 wagons. Barnwood Driver H Fluck reported 60 minutes were lost on the journey from Evesham to Redditch, due to the engine slipping. The leading sanders did not work because of wet sand in the traps. An assisting engine was provided from Redditch to Barnt Green.

Just for a change the empty, rather than the full, Bromford Bridge tanks had problems soon after midnight on 2 December. The working was the previous night's 8.50pm to Fawley and was hauled by 4F 44004. This must already have been deputising for the normal 9F. The 4F was short of steam – not surprising with a dirty fire, inferior fuel and a blocked ashpan. According to the report, prior to this working the loco arrived at Saltley at 5.20pm and left at 7.5pm, so the chances are, apart from coaling up, it did not receive any servicing in the short time it was there. Barnwood sent it home light engine. There is no indication of what loco took over the train.

A visit to Cheltenham Malvern Road on 2 December found three of Barnwood's own 4Fs recently stored in the shed – 43887, 44209 and 44296. All went back into traffic either in 1962 or 1963. Also there on 2, 3 and 4 December was 73093, likely to have been the engine that brought the 4Fs over from Gloucester.

Mention of Malvern Road brings in a loco type not featured to date in the casualty reports – the 'Jinty' 0-6-0T, of which Gloucester had several examples. (Midland men called them 'Jockos' not 'Jinties' though the latter was the term commonly used by enthusiasts. The Deeley 0-4-0T at Gloucester for docks shunting were, however, known by Midland men as 'Jinties'.) 47417 came to grief on Wednesday 20 December while shunting at Cheltenham High Street Goods. Well into the 1950s Cheltenham High Street had footplatemen stationed there for this and maybe other duties, who came under Barnwood's wing. Fred Cole confirmed there were three sets when his time started at Barnwood, though unfortunately they were later made redundant. This was probably a throwback to the days when the MSWJR had a functioning loco shed at High Street. It was absorbed by the GWR in 1923, but the LMS shunting engine – usually an 0-6-0T – stabled there until it closed in the mid-1930s. Anyway, by the date of 47417's problem, the loco was stabled at and manned by Cheltenham Malvern Road shed. Shortage of water in the boiler resulted in a fused safety plug. There was no mechanical reason for this – apparently the very inexperienced fireman filled the loco's tanks but neglected to put the injectors on to fill the boiler as instructed by the driver who was having his break in the cabin. The old-hand driver had an uncomfortable interview at Swindon to explain why this casualty occurred. 47417 had been pretty active, with an estimated 55,052 miles done since a general overhaul at Derby Works dated 11 April 1958. 47417's replacement on the High Street duty was 3F 43754, seen later that day on Malvern Road shed.

Despite closing around seventy years ago, the three-road MSWJR loco shed still exists at Cheltenham High Street, and is now used for retail purposes. And Cheltenham Malvern Road shed buildings also survive, adapted for use by a builder's merchant.

Below: A Barnwood 'Jinty' 0-6-0T was outstationed at Cheltenham Malvern Road – a sub-shed of 85B Gloucester Horton Road – to shunt the goods yard at Cheltenham High Street. 47417 posing here outside Malvern Road depot, was unfortunate enough to suffer a fused safety plug while performing the duty on 20 December 1961. 47417 moved to Barnwood in December 1955 because it was vacuum-fitted, as shown here, enabling it to perform as station pilot at Cheltenham Lansdown station, in addition to its duties at High Street. *R Stanton*

OTHER CASUALTY REPORTS October-December 1961

Date	Loco	Home shed	Train	Problem	Place	Delay (mins)	Notes
16/10/61	92078	18A	12.38am Millbrook-Bromford Bridge class D	SOS	GLOS	69	1
18/10/61	44179	21A	6.30pm WWH-Bristol class C	SOS	ER.		2
18/10/61	44184	21A	8.55pm Avonmouth-WWH Banana train	SOS	CHEL		3
24/10/61	73028	82E	4.20pm Bristol-Gloucester class B	CWA	ER.		
30/10/61	46131	41C	7.10pm Newcastle-Bristol class A	SOS	ER		4
02/11/61	44811	15C	12.43pm Newcastle-Bristol class A	SOS	ER	35	4
04/11/61	45253	21A	8.55pm Branston-Newport minerals	INJ	CHEL		2
23/11/61	45654	41C	10.30am Manchester-Bournemouth class A	SOS	ER		5
26/11/61	45658	55A	12.15pm York-Bristol class A	PRI	ER	7	
02/12/61	44112	21A	3.25pm WWH-Westerleigh freight	SOS	ER		3
24/12/61	48197	18A	7.0am WWH-Westerleigh freight	SOS	ER		3

Notes
1   Loco to Barnwood for examination, dirty fire, exhaust injector not working
2   Engine light to Saltley for attention
3   Engine to Barnwood for examination
4   Loco off train at Gloucester, to Barnwood shed
5   Loco light to home depot for attention

Below: 46131 *The Royal Warwickshire Regiment* was short of steam on 30 October 1961, as noted above, but has steam to spare in this portrait at Hatherley depicting it heading north on 1N61 7.45am Paignton-Newcastle, a summer Saturday train.   *R Stanton*

CHAPTER FIVE

REPORTS FOR 1962 AND 1963

Officially it was hoped all regularly timetabled class 'A'/class 'I' express passenger trains on the Birmingham-Bristol line would have Type 4 diesel haulage from January 1962. This did not happen, and even by June 1962 just 80 % of these services were claimed to be diesel worked.

The Fawley-Bromford Bridge oil trains were diverted to run over the Didcot, Newbury and Southampton line in 1962 because, it was stated, of heavy delays encountered in the Bristol area. But 'Esso' tank trains still ran through Gloucester going from Avonmouth to Bromford Bridge.

On the Dursley branch, the last passenger trains ran on 10 September 1962, hauled by Ivatt 2-6-0 46526. The freight service however continued and Dursley loco shed stayed open.

With a couple of exceptions, the 1962 reports are not available, presumably having been destroyed.

On 14 November 1962 92008 of Saltley was heading the 1.0pm Hallen Marsh-Widnes tanks. Barnwood Driver N R Sutton, with Fireman R Sanders, reported: 'I was stopping at Ashchurch for signals when the right boiler gauge frame packing nut blew out filling the cab with steam and water. We were hauled inside the Stores at Ashchurch and after making several attempts to put the fire out, we got soaked to the skin, my fireman ripped his trousers from top of the leg to the bottom and the stitching came away from my shoes. We had to use several buckets of salt out of a shed in our attempt to put the fire out.' The train eventually headed north behind 4F 44148 having been delayed by two hours 28 minutes.

The salt mentioned was for use by the permanent way department, while the Stores were the old Midland Railway provender stores, one of only two owned by that company, which had provided feed for the Railway's thousands of horses.

The four-storey building was a prominent landmark at Ashchurch, which had a thriving railway community for years – over a hundred people were employed there around 1910. The incident with 92008 was shortly after the formation of the Dowty Railway Preservation Society, which came to use the store's sidings. By this time the stores building was no longer in railway use and although a fire in later years led to its reduction to two storeys, it still remains a landmark today.

Back to the tale of the crew of 92008, who put in claims for their ruined items. The driver valued replacement shoes at 46/6d, while the fireman was looking for 50/- for replacement trousers. But after a lot of correspondence, with management saying that compensation could only be based on the value of the articles at the time, Driver Sutton received 23/- and Fireman Sanders 20/- in June 1963.

Observation at Ashchurch a few days later, on 23 November, saw an interesting combination on the 8.40am Worcester-Gloucester local passenger, 0-6-0PT 4614 assisting a failed D140.

A loco casualty which suffered terminally by coming to Barnwood for repair was 'Castle' 5049 *Earl of Plymouth* of 82B St Philip's Marsh – no stranger here having had a previous hot box attended to in October 1961 as already related. It got another hot box, detected at Bristol Temple Meads on 17 December 1962 whilst in charge of the 7.50pm Avonmouth-Tavistock Junction, 4C25. But 5049 never went back into traffic, as while standing at Barnwood awaiting attention, it sustained a cracked cylinder during the severe freezing weather. 5049 was withdrawn in March 1963.

JANUARY – JUNE 1963

'...It was impossible to maintain a full head of steam with a fire in this very dirty condition. Two foot of solid clinker in firebox – smokebox full to 6" above bar.'

1963 started with the freezing weather which hit the country at Christmas 1962 and was set to continue well into March – it was the worst winter since 1947 with heavy falls of snow. Services were disrupted for weeks with some trains cancelled and others running very late. Lots of diesels froze up which meant steam locos often substituted on services throughout 1963 to cover for diesel shortages. And diesel/steam combinations became familiar, with the steam loco providing train heating. Late running played havoc with booked duties for both train crews and engines, causing the appearance of many locos from depots not normally associated with the area. One such on 30 December 1962 was 6B Mold Junction's 45438 on the down 'Devonian' running 60 minutes late at Cheltenham.

'Peak' D39 was an outright failure on 4 January 1963 with the 10-coach 1V38 Sheffield-Bristol express, and left Cheltenham 149 minutes late being towed by Horton Road's 0-6-0PT 9471. The extent of diesel failure and non-availability can be gauged by the fact that of eighteen cross-country expresses seen at Cheltenham on 4 January, only five had diesel power – including D39 – , one was doubleheaded by diesel and steam, with the remaining twelve being steam worked.

It wasn't just diesels failing that day. Leeds Holbeck 'Jubilee' 45568 *Western Australia* had the 5.37pm Bristol-York express, 1N87, but was in trouble soon after departure and only got as far as Gloucester, losing 19 minutes. It was short of steam, with a full ashpan preventing air getting to the fire which was very dirty, not helped by poor quality coal. It all smacks of a lack of servicing; perhaps being turned off Barrow Road shed as a hasty replacement for a failure soon after arriving off its previous working. 73019 took the train north from Eastgate, being observed at Cheltenham 68 minutes behind schedule.

Long-distance passenger trains seen on 4 January – excluding the aforementioned 1V38 – were up to 70 minutes adrift of right time, so star turn of the day must be awarded to the down 'Devonian', just

Above: One of Holbeck's finest, 'Jubilee' 45597 *Barbados,* heads a down express at Hatherley. It did however spend virtually the whole of August 1960 laid up at Barnwood for hot box repairs. On another occasion – a very wintry 4 January 1963 when services were in chaos due to bad weather and engine failures – 45597 substituted for a diesel on the down *'Devonian'* and was running a mere 5 minutes late at Cheltenham, indicating it was in exemplary condition.

R Stanton

five minutes late at Cheltenham with Holbeck's 45597 Barbados, *obviously in much better fettle than shedmate 45568. The* 'Devonian' *was timetabled for a non-stop run of one hour 52 minutes over the 89 miles from Birmingham to Bristol.*

A few days later, on 8 January, another Holbeck 'Jubilee' – 45562 *Alberta* – was sent home light engine from Gloucester after failing with superheater elements blowing while working the 3.0am Leeds-Bristol. Meanwhile 45568 had recovered as it was on 1N84, the 1.40pm Bristol-York express that day, almost on time.

Stoke depot, 5D, was not often represented in the area, so when their 4F 44536 appeared on the 4.20pm express beer train to Bristol on 7 January, it had no doubt been 'borrowed' for the duty by Burton shed, who normally supplied their own loco for the train – maybe a 'Crab' or one of their allocation of somewhat rundown 'Jubilees'. 44536 was no stranger to the route, though, it spent many years as a Bristol Barrow Road engine up to October 1959 before being transferred to Stoke. Barnwood Driver Lionel Lord reported 'While running between Pirton Sidings and Eckington I could smell something burning and suspecting a hot axle box, I reduced speed to Ashchurch. Having to stop at Ashchurch for traffic, I examined the engine and found right driving axle box hot ... I informed Control from Ashchurch at 8.35pm fresh engine was required from Gloucester.' The replacement was 44213. 44536 spent quite a few weeks at Barnwood awaiting repair, still being there on 21 March. The fireman on 44536 was Mike Randall who was the regular mate of Lionel Lord for a couple of years. He recalls this incident and said that Lionel always drove at high speed. On another occasion, they got a hot box at Wickwar tunnel on an up freight with an 8F. Mike

remembers they came out of the tunnel 'faster than a cork out of a bottle'. He says it took some doing to get a hot box on an 8F.

A rare loco for the area was 45416 of Bedford depot. It expired at Cheltenham at 2.15am on 11 January whilst working the 8.25pm Templecombe-Derby class 'C' perishables. Bath Green Park men were in charge. The loco went to Barnwood for attention where the foreman commented 'Fire very dirty, heavy clinker. This was never a clean fire from Bath. Also poor quality coal.' Yet another example of lack of servicing. But after attending to it, Barnwood felt confident enough to put 45416 on the diesel rostered 1N61, 3.15pm Cardiff-Leeds, from Gloucester later that day – and it swept through Cheltenham non-stop this time.

Earlier on that winter's day the 1.40pm Bristol-York class 'A' failed at Gloucester with Burton shed's 45649 *Hawkins* having a broken blower pipe. Nos. 44965 and 44777 worked forward.

It was dieselisation which caused sheds such as 17B Burton to have an unaccustomed allocation of types like 'Jubilees', receiving no less than eighteen of the class in November 1960, with a few more at a later date. 16D Annesley did even better, not only receiving some 'Royal Scot' 4-6-0s as diesels spread throughout BR, but also 'Britannia' 4-6-2s. 46126 *Royal Army Service Corps* of 16D headed the 3.0am Leeds-Bristol express passenger, 1V28, on 14 January. But there were problems with the left-hand injector so the 'Scot' came off at Gloucester, with a minimal delay of 7 minutes down to the engine. And just to illustrate that it was not always steam replacing diesel, the fresh engine on the train this time was D62.

46126 was another loco which had spent time at Rugby Test Plant, during 1957.

On 15 January, D25 expired with the down 'Devonian' and suffered the ignominy of being towed through Cheltenham behind 0-6-0PT 4697.

9F replaced 9F later that day on the 2.30pm Avonmouth-Bromford Bridge tanks. Saltley's 92107 had suspected superheater elements blowing and was sent home light engine, while shedmate 92152 worked the train after a 65 minute delay.

'Britannias' appeared fairly regularly in this period due to the shortage of diesel power; 70049 *Solway Firth* of 16D was on a down freight on 20 January, while 21D Aston's 70029 *Shooting Star* had the 7.35am Bristol-Bradford express on 22 January. The crew complained of condensed steam in the cab, so 73094 worked the train north. But no problems could be found on examination at Barnwood and 70029 provided superpower for the 4.43pm stopping passenger to Birmingham that day.

When 70029 failed, the delay was noted as just 15 minutes. This was minor compared with some other trains on 22 January: the down Newcastle-Cardiff, 1V70, was 161 minutes late at Cheltenham with 45614 Leeward Islands; 1V45 Newcastle-Bristol was 112 minutes late with 45186; and 1V46 Sheffield-Bristol was 138 late with D153. The week of 21-26 January was particularly bad, with a number of passenger trains cancelled on one or more days – something virtually unheard of on BR in that era, except in the most extenuating circumstances.

24 January saw an unusual combination on the 7.55am Swansea to Newcastle express – diesel D155 and BR Standard class 2 2-6-0 78009 of 85A Worcester, presumably the latter was providing steam heating. The corresponding working that day, the 8.20am Newcastle-Swansea, was hauled from Derby to Birmingham by no less than 34A King's Cross A3 Pacific 60039 Sandwich – that would have been an interesting sighting at Gloucester! But the train changed locos at New Street and was running 147 minutes late when seen at Cheltenham with a rather mundane 'Black Five', 45280.

Another 'Britannia' appeared on 25 January, Aston's 70025 Western Star, working a down late evening express. It returned north next day on empty coaching stock.

Barnwood men had a good run with Saltley 4F 44160 on 26 January. They went non-stop from Cheltenham to Westerleigh on the 6.40am class 'E' ex-Washwood Heath and non-stop light engine to Barnwood. But while disposing of the loco on shed, they noticed the left leading engine axle was hot, so a repair card was completed. The leading axle journals were turned and the boxes reconditioned.

Around a year earlier, 15 January 1962 to be precise, 44160 was a very unusual visitor observed by the author on the Cheltenham-Kingham line; unfortunately he has no record of what it was working, but vaguely recalls it being light engine.

Barnwood's own 73096 worked the 2.55am Cardiff-Woodford Halse freight on 28 January but failed at Severn Tunnel Junction with a Cardiff crew in charge. There was a delay of 70 minutes and 3840 took the train on. The problem on 73096 was the regulator valve blowing through. It was later taken out and lapped in and the regulator gland repacked. The Barnwood report showed the loco had covered 53,400 miles since its last general overhaul, at Doncaster, dated 24 March 1961.

Another Aston loco, 'Black Five' 45114, delayed the 2.10pm class 'C' Penzance-Crewe fast goods by 43 minutes on 29 January. It was short of steam which the Bristol St. Philips Marsh driver blamed on the smokebox drawing air and the exhaust injector not working. But Barnwood could not find any defects. Nevertheless, 44943 took the working north. 45114 later successfully worked the 5.40pm Gloucester-Bristol stopper and 10.5pm return. Barnwood stated 'It can only be assumed that the St. Philips Marsh fireman was unfamiliar with the class of engine and mismanaged the fire.' The Penzance-Crewe was normally worked by a GWR loco type and, in the normal course of events, crews from 82B probably did not get much exposure to 'Black Fives'.

Burton's 45585 *Hyderabad* went down with the 12.15pm Newcastle-Bristol express on 2 February, noted 86 minutes late at Cheltenham. But it still had plenty of time to be serviced at Barrow Road before returning on the 1N68 Plymouth-Bradford, 4.45pm ex Bristol express passenger on Sunday 3 February. However it was in poor condition again, short of steam, losing 15 minutes between Gloucester and Cheltenham, more than doubling the scheduled running time. Classmate 45597 *Barbados* was on the following 5.55pm Gloucester-Birmingham stopper and went to Cheltenham High Street to assist 45585 forward. This left the stopping passenger without a loco at Lansdown station; eventually 'Britannia' 70018 *Flying Dutchman* – superpower again – arrived from Gloucester and took the train out 62 minutes late. 70018, of 5A Crewe North, was ex-works and had arrived in the area on a down stopper on 1 February. (Yet another 'Britannia' 70050 *Firth of Clyde*, also of 5A, was on an up fitted freight on 18 February, making at least five of the class on the line in the space of a month.)

Sundays were often notable for the number of unbalanced freight workings, usually coal trains, which came from the north, resulting in strings of engines heading back light. 3 February was no exception, with no less than 13 locos seen on the up, including five locos coupled together, the maximum permitted.

The down 'Devonian' did not fare too well on 5 February. This time it had Holbeck's 45605 *Cyprus*, with a Barrow Road crew from Birmingham New Street. They lost 25 minutes to Gloucester, due to a dirty fire and inferior small coal. 44852 took the train to Bristol, though it is worth repeating this train should have been a diesel throughout.

One of Bristol's remaining 'Jubilees' 45682 *Trafalgar* was working in the opposite direction on the 2.15pm Bradford class 'A' on a snowy 10 February, but gave up at Gloucester, with a full ashpan, smokebox full of char and a dirty fire, according to G Blunt, foreman at Barnwood.

A loco casualty at Ashchurch was the first preserved steam locomotive to arrive at the Dowty Railway Preservation Society site. Avonside 0-4-0T 'Cadbury No.1' came by rail from the famous chocolate factory at Bournville on 16 February 1963 hauled by a 4F and got a hot box on the journey (reminiscent of 41537, another 0-4-0T, two years earlier).

Saltley's 92136 found itself on glamorous work on 20 February

The harsh winter of 1962/63 played havoc not only with services, but also with diesels, just as they were consolidating their grip on express passenger and freight turns on the Birmingham-Bristol line. On 20 February 1963, D44 became a casualty at Cheltenham High Street with the up 'Devonian'. 92136, waiting in the loop on a freight, was called upon to provide assistance and these three pictures, taken from High Street signal box, show the sequence of events: the two locos side by side; 92136 just starting away, complete with class 'A' headcode; dead D44 being towed behind a very leaky 92136 – hope they made it!
R Stanton

Above: Gloucester was required to supply a class 5 loco as station pilot at Eastgate in later steam days so that a suitable engine was quickly available to take over from failed motive power on expresses. A number of BR Standard 5s were allocated for this duty and other tasks; one of the Gloucester allocation, 73092, is seen passing the Midland Railway signal box at Bredon with a class 'B' passenger service. One casualty report from 20 March 1963 has 73092 taking a northbound working in place of 'Jubilee' 45557 *New Brunswick* which was declared a failure after arrival at Gloucester with the down leg of its diagram – though it should be mentioned that 45557 was already substituting for a diesel!

when D44 failed at Cheltenham High Street with the up 'Devonian' and 92136 was summoned off an up freight to provide assistance. Pictures of this event taken by Robin Stanton who was at the adjacent signal box show an extremely leaky 92136, not inspiring much confidence in its ability to shift this important train north at anything like express speed!

And 92136 was indeed in trouble on 23 February, but with a hot right driving axlebox rather than a steam deficiency. This happened at Spetchley while hauling the 6.30pm Avonmouth-Bromford Bridge tanks. Driver George Jones of Barnwood related '...informed Control we required a fresh engine and proceeded to Bromsgrove at reduced speed and put engine on Bromsgrove shed for box to cool down. At this stage the box was on fire, which I ventured to put out.'

92136 had recovered by 1 March when it worked the 10.40am Bromford Bridge-Avonmouth, but managed to fail from shortage of steam this time. A very dirty fire, heavy clinker and poor quality fuel all contributed to its downfall in the Gloucester area. The train was delayed by 20 minutes and 'Black Five' 44813 took over for the rest of the journey.

Next day, the tanker train in the up direction, 10.40am Avonmouth-Bromford Bridge, suffered 63 minutes delay when Wellingborough's 92132 had problems with tubes leaking at the firebox end.

Not a failure concerning Barnwood, but on Sunday 3 March, Hymek Diesel D7006 expired while working the 9.30am Paddington-Cheltenham passenger, which passed Lansdown Junction two hours late with 4914 Cranmore Hall *tender-first towing the diesel and train.*

On 8 March, the up Esso was worked by an Eastern Region 9F, 92144 of 34E Peterborough New England depot, slightly more exotic

than the usual 9Fs from Bristol, Saltley or even Wellingborough!

Barnwood's BR Standards were not immune from problems. 73094 had the 6.46am Eastgate-Sheffield passenger on 26 February with a lightweight load of 184 tons. It was nearing the end of the turn when it failed at Wingfield with a Derby crew, Driver Burbage and Fireman A Brown. The problem is shown as 'Jet pipe on smokebox defective.' Delay was 55 minutes. The report shows that 73094 had covered 62,757 miles since an intermediate shop repair at Doncaster dated 3 November 1960.

73092 was also up in the East Midlands, as next day it worked the 7.35am Nottingham-Bristol passenger. It was in charge of a Saltley driver and fireman when it expired at Gloucester short of steam, causing a 35 minute delay. Barnwood serviced the loco and used it on the late afternoon stopper to Birmingham.

The same Saltley crew had more problems on Tuesday 5 March with the 12.15pm Newcastle-Bristol express, hauled by 45659 *Drake*. The engine had worked from York, but was not steaming well, so came off at Gloucester, replaced by 44966 which had been on the following stopping passenger. Delay due to the engine is shown as 40 minutes, so other problems must account for the fact that the express was 71 minutes behind schedule when noted at Cheltenham. Examination of Holbeck's 45659 at Barnwood found 'Fire very dirty, heavily clinkered, ashpan full to bars and smokebox full to top of blastpipe.' No wonder the loco failed, it illustrates clearly the problems that footplate crews had to contend with, often day in, day out.

On 10 March, another Holbeck 'Jubilee' 45569 *Tasmania* was on IN78, the 2.15pm Bristol-Bradford express, but only got to Gloucester. Same story – engine not steaming. 45569 was sent light to Leeds for attention. When observed at Cheltenham the train was 36 minutes late, doubleheaded by 'Black Fives' 44941 (which had piloted 45569 from Bristol) and 44963.

Two days later, 44941 had a rather different task, hauling recently withdrawn 'Castle' 5049 Earl of Plymouth from Barnwood depot to Cheltenham Malvern Road shed, where 5049 remained for several months before going for scrap at Cashmores.

Engines were capable of completing turns almost to schedule despite evident lack of servicing. One such occasion was 18 March with 45088 on the 7.10pm Birmingham New Street-Gloucester stopping passenger. This was a pretty easy turn for a 'Black Five' and it was seen at Cheltenham just six minutes behind time. However here is what the Worcester men had to contend with, as reported by the foreman at Barnwood: 'Fire very dirty. Clinker on bars a foot thick, smokebox full to top of blast pipe, poor quality coal. Smoke plate and fire hole ring in back of firebox.'

A failure earlier in the month, 45659 was on 1V67, the 1.8pm Leeds-Cardiff, on this day and running well.

However bad the state of 45088, 17B Burton's 45557 *New Brunswick* was even worse on 20 March and on a far heavier train, the 8.20am Newcastle-Swansea express loaded to 385 tons. The train was 80 minutes late at Cheltenham and the whole of the delay was booked against the engine. Barnwood Driver Bill Dipple and Fireman D Norman had a struggle, reporting: '...It was impossible to maintain a full head of steam with a fire in this very dirty condition.' Even the shift foreman seems to have been shocked by the conditions he saw, noting: 'Two foot of solid clinker in firebox – smokebox full to 6" above bar.' The loco had worked through from York. No-one could have blamed any of the crews involved if they had demanded a fresh engine en route. 45557's return working on this diesel diagram, the 3.15pm Cardiff-Leeds, was taken by Barnwood's 73092.

Steam was still covering for diesel non-availability on some class 1 trains in May and 1E64, the 8.40am Bristol-Sheffield, was often steam-hauled. Barrow Road's 45690 *Leander* had the duty on 9 May, but gave up at Gloucester with a lubricator problem. 73019 worked forward after a 30 minute delay. A temporary repair was effected and 45690 returned light to Bristol.

Another 82E Barrow Road engine had problems on 22 May. 92000 was in charge of the previous night's Esso tank train, 11.30pm ex Avonmouth to Bromford Bridge. Barnwood Driver J Lees and Fireman Staples completed the trip and put the engine on Saltley depot with a repair card requesting attention to the defective exhaust injector, which had caused them to lose time. The next day, they were on the working again as was 92000, but this time failed it at Gloucester because of the injector and the crew complained that nothing had been done to rectify the problem. 92118 took over, but the delay to this important train amounted to 90 minutes.

Later on 23 May, an interesting special ran from Gloucester Old Yard. It was a train of nine horseboxes, load 118 tons, destined for hunting country, Melton Mowbray. 44092 of Saltley had the task of getting it there. With such a lightweight load, it can be surmised that Barnwood Driver R B Roberts and Fireman Smith would enjoy a good gallop with their steed. Perhaps they did, because the left big end got hot, discovered at the first booked stop, Bromsgrove. A fresh loco was needed and

Bromsgrove turned out one of the bankers, 0-6-0PT 9401, which escaped the confines of its duties on the Lickey Incline and took the train to Duddeston Road, Birmingham, from where no doubt a loco came off Saltley depot to take over. Delay at Bromsgrove was one hour, plenty of time to feed and water the horses in their boxes. 44092 was laid up, with wheels sent to Derby for attention to the journals. It had only recently received an intermediate overhaul there, dated 29 March 1963.

In contrast 92157 had not received a works overhaul since emerging brand-new from Crewe on 20 November 1957. Its record indicates that it had amassed 157,700 miles since then up to 23 May 1963. Driver Lees and Fireman Staples had another rough trip on the 11.30pm from Avonmouth, load 25 tanks, due to 92157 having defective injectors. The live steam injector cone had come unscrewed and the clack of the exhaust steam injector was off its seating due to scale. Consequently they lost 70 minutes on the journey to Bromford Bridge and once again had to book defects on arrival at Saltley loco on 24 May.

Barnwood 8F 48420 suffered a collapsed spring on 12 June while working the 9.40pm Washwood Heath-Westerleigh class 9 freight. This happened at Camp Hill and 44180 took over after 45 minutes. 48420 had done an estimated 49,000 miles since its last shop visit for a heavy intermediate overhaul at Crewe dated 28 August 1961.

A seaside excursion, 1X71, on 14 June brought Aston's 'Britannia' 70047 to Weston-Super-Mare from Stechford, Birmingham. Actually the loco had come off Rugby shed at 6.13am for the working, having laid over there since 11.35am on 12 June. Rugby supplied it with 5 tons of Welbeck Colliery grade 2 coal. All went well until the return working, 5.55pm from Weston, which was a very heavy train, 15 coaches, justifying its class 7 power. Barnwood received a message that the Bristol driver was asking for a fresh engine as he was running out of coal. 45199 went off shed at 7.15pm to replace 70047, which came to Barnwood where it was found to have just 40cwts of coal left on the tender. 70047 steamed north next day on a fitted freight, 5M27, 7.0pm Bristol St Philips to Derby.

Seeing a 'Britannia' at Gloucester was reasonably exciting, but on 27 June, this was easily surpassed with reportedly the first ever recorded sighting of an Eastern Region class 'K1', 36C Frodingham's 62035 on a down fitted freight. It returned next day on a freight headed for Woodford Halse (which would suggest it traversed the Cheltenham-Honeybourne line). Engines from Frodingham were not at all common – though the previous day saw another one, in the rather more mundane form of 'WD' 90439, on a special train of molasses.

90439 was purloined for local workings, as it had the 2.5am Washwood Heath-Westerleigh class 8 freight on 28 June. It became a complete failure at Naas Crossing, Gloucester, when both injectors packed up, necessitating the fire to be drawn, fairly dramatic stuff. A small piece of coal had got lodged in the delivery cone of the right hand injector, so this was subsequently cleaned and refitted. The clack on the left hand injector was blowing through badly and after inspection this was ground in at Barnwood. The depot master at Frodingham forwarded a very neat, precise engine history which is copied here. 90439 returned north light engine on 29 June.

Summer Saturdays still saw a fair amount of booked steam on extra passenger workings. 45561 *Saskatchewan* failed at Cheltenham on 1V38, 7.30am Newcastle-Paignton, on 29 June. While it went to Barnwood light engine, 73091 came up from Gloucester to take over the train, which left Cheltenham 76 minutes late.

JULY – DECEMBER 1963

'...I consider the driver and fireman must be held responsible for this incident.'

For the last period in our survey it is noticeable that there are few reports for July, August and September, though the last three months of October, November and December seem to be more in the usual pattern. The reason is not known – maybe a load of reports got thrown out. Perhaps it reflects more diesel working, whose failures would not necessarily involve Barnwood. Perhaps there was less paper shuffling with the rundown of steam. Anyway there are still some interesting items to browse through.

Not in the reports, but observed by the author, was the up 'Devonian' on 16 July, worked by 73091, to the rescue once again, this time of a failed 'Peak' diesel.

8F 48133 was on the 2.10pm Washwood Heath-Stoke Gifford class 8 freight on 17 July. Aboard were Barnwood's Driver E. Badham and Fireman Cottrell. The driver recorded 'Engine became disabled at Frocester at 8.30pm due to left cylinder becoming fractured. Rules carried out and line cleared at 9.30pm.' The casualty report states 'Lock nut of left piston became detached causing piston head to force nut against the bottom of cylinder cover, resulting in cylinder and front cover fracturing.' The loco was proposed for a special visit to shops and was seen travelling north on 29 July in the company of 45246 and 43122.

There are no reports in this period about the Esso trains, although 4F 44577 of 16A Nottingham was seen on the down tanks on 25 July, almost certainly substituting for a failed 9F, while an Esso working on 17 August had no less than GW 4-6-0 5972 *Olton Hall*.

Former Crosti 9F 92028 of Saltley had a turn on 7 September which involved taking a special freight from Stanton Gate in Derbyshire to Pilning. It came off Toton shed for this duty, having had its fire dropped and the ashpan and smokebox cleaned during a 12-hour layover on 6 September. But it managed to fail at Cheltenham short of steam, delaying the train by 66 minutes.

Fairly unusual on 7 September was the use of 9Fs – Saltley's 92138 and 15B Kettering's 92080 – on stopping passenger trains through Cheltenham. 92138 is thought to have been the first 9F seen at Gloucester on a passenger train on the Birmingham to Bristol line – a Summer Saturday extra on 27 July 1957.

Barnwood supplied 8F 48167 on 9 September as a replacement at Gloucester for hot box victim 6950 *Kingsthorpe Hall* of 88A Cardiff Canton. The train was the 7.10am Severn Tunnel Junction to Lawley Street freight, crewed by Severn Tunnel men.

BRITISH TRANSPORT COMMISSION   B.R. 1/7

BRITISH RAILWAYS

8 JUL 1963 W.R. GLOUCESTER

M.P. _____ DEPARTMENT

E. _____ REGION

FRODINGHAM _____ STATION

Your Reference

5th July 1963.

Motive Power Depot,
BARNWOOD.

      Casualty to Loco' 90439.
         28.6.63.

         Referring to your initial report, I give below the history details required:-

Last Shop Repair ... 5/62 at Gorton
   Estimated mileage since ... 27,000
Last 'X' Exam. ... 22.6.63 at Frodingham
   Working days since ... 6.
Last Weekly Exam. ... 22.6.63 at Frod.
   By G.Mullen, 5474, Fitter.
Last Periodical Exam. ... 11.4.63.
   Item No.4344 at Frodingham.
Est. period/mileage ... 3,000.
Defective part last exam. ... 11.5.63
   at Frodingham by R.Barnard, 5476,
   Fitter.
Previous bookings ... Nil.

         DEPOT MASTER.

9 September was the first day of the winter timetable and while there was now less steam substitution for diesel than in the early part of the year, most days saw some such workings. This day the down 'Devonian' was steam with 'Jubilee' 45675 Hardy of Holbeck, keeping up the old traditions by working through from Leeds, albeit 23 minutes late. The diagrammed diesel return working, 1N87, 5.40pm Bristol-York, was hauled by Barrow Road's 45690 Leander.

On 24 September, Barnwood provided its own 8F 48463 to take over from Saltley 9F 92152 which failed at Stonehouse. The live steam injector clack was blowing, unable to draw water, while the exhaust injector was not working on exhaust steam. A big delay, 230 minutes, held up the working, the 10.58am Washwood Heath-Stoke Gifford. 92152 had covered just 7,000 miles since a casual shop repair at Crewe.

Next day, 48463 itself needed rescuing. It had the 6.25pm Stoke Gifford-Washwood Heath with Barrow Road Driver Jaques and Fireman Pameo, load 57. Failure occurred at Gloucester due to a steam blow from the front end. This was due to a temporary repair to a pipe having fractured again. 44226 took the freight on to Birmingham.

Above: 44123, a Barnwood engine, is seen passing Tramway Junction box, Gloucester, on 23 March 1963. A few months later, on 2 October, it suffered a fused firebox safety plug at Hempstead, necessitating extensive repairs. The loco was being crewed by Gloucester Horton Road men at the time of the failure.
B W L Brooksbank

Another Barnwood engine was in trouble – big trouble – on 7 October. The 6.45am trip working to Hempstead sidings in Gloucester Docks, load 16, sounds like a doddle for 4F 44123. But, perhaps due to a shortage of crews at Barnwood, it was worked by Horton Road men. Western men and '4Fs' had been in trouble before. And it happened again this time. Due to a shortage of water in the boiler the engine suffered a fused firebox safety plug at Hempstead. Blame was placed squarely on the engine crew. The Barnwood shed master commented 'I am satisfied that the engine was in good condition when it left shed, and in view of the gauge frames being found to be in good order following the mishap, and the safety plugs having been examined only twelve days before, I consider the driver and fireman must be held responsible for this incident.'

The loco needed extensive repair. Apart from a new safety plug and attention to a leaking tube plate seam, 40 roof stays were changed and 12 superheater flues expanded. 44123's last shop repair, a general, had been at Derby, on 1 September 1961, since when it had worked 40,016 miles.

Another Barnwood 4F, 44045, had a mechanical failure the very same day, though details are sketchy. It had to go on Bromsgrove shed for attention while working the 3.15pm Barnwood-Lawley Street goods. This loco had a general at Derby dated 15 June 1961, having since covered 46,926 miles, so had been a bit more active than shedmate 44123.

October 7 was a busy day for failures. Saltley 8F 48339 had Barnwood Driver F A Rigby and Fireman B Fernihough on the 6.25pm Stoke Gifford-Washwood Heath freight, load 54. At Hatherley, the driver slowed for a signal check, but while closing the regulator the engine gained power. Presumably prompt action using reverser and brake stopped it before running through the signal. The driver stated that the regulator valve had become uncoupled, so he failed the engine. He took the train to Cheltenham High Street where it was terminated and the loco ran back light to Barnwood. Examination there provided the following information "The regulator valve had not become uncoupled but would not shut off because of play in the drive from rod to valve. If this valve had been slammed shut it is quite likely that it would have been thrown shut.'

Barnwood's 73068 suffered a mechanical defect two days later while working the 10.5pm stopping passenger from Bristol to Gloucester, 2H74. Driver F Salt, with Fireman R Hawkes, stated 'On running into Coaley station, the right hand side eccentric rod broke causing engine failure. Had to get fresh engine from Gloucester, delay 101 minutes.' The incident was investigated by the Senior Technical Assistant (Locomotives) who was based at Bristol. He reported 'Eccentric return arm fractured due, I suspect, to initial damage caused by a side blow bending the crank. The bent crank caused the arm to whip excessively and ultimately fracture mid-way along the arm. The crank pin radial ball was also damaged due to excessive alignment of the crank. I suspect the damage was sustained whilst on shed during a movement incurring a minor collision.' Despite some follow-up, it appears that the culprit was not traced.

An earlier stopper that evening, 2M74, Bristol-Birmingham, also appears to have had engine problems, as it was seen at Cheltenham 112 minutes late, doubleheaded by 44660 and 73021.

21A 9F 92157 had charge of a special freight from Clifton to Oldbury on 17 October with Bristol Bath Road men aboard. But it failed at Gloucester as both injectors were defective. Changing engines made for a 31 minute delay, but at least the Western men had a Western engine as replacement, 6925 *Hackness Hall*.

92157 had still not had a works visit since being new in 1957; its recorded mileage was now 167,000, compared with 157,700 in a previously mentioned casualty report of May 1963.

The 5.0pm Milford Haven-Coleshill tank train, 4M24, was rostered for haulage by a 'Western' diesel hydraulic, but availability of that type was very poor at the time. On 21 October 86A Newport Ebbw Junction provided its 9F 92241 for the working, but a mudhole joint blew out, so the loco was failed at Barnwood, where 'Jubilee' 45674 *Duncan* took over after a 60 minute hold-up.

Another potential failure at Barnwood two days later was 44263 on the 2.20pm Westerleigh-Washwood Heath. The Bristol driver who brought the train to Engine Shed Junction, Gloucester, was concerned about the left driving axle running warm, so the relieving Barnwood men called for a fitter. He examined the problem and declared the engine fit to carry on. So it did, but on arrival at Bromsgrove, metal was running from the axlebox. A passing 'Black Five' 44919 worked the train forward, while 44263 trundled off, very slowly, light engine to Saltley.

If the statistics supplied by Saltley can be believed, 44263 had done 81,000 miles since a general overhaul at Derby dated 1 December 1961. Even for Saltley, whose engines had a reputation for turning up in the most unlikely and far-flung parts of BR, this seems extremely high and surely there was a clerical error here!

6816 *Frankton Grange* was entrusted with the 11.25pm Bristol-Crewe parcels, 3M00, on 28 October. By the time it got to Gloucester Eastgate it was short of steam, so retired to Barnwood depot. The train left with 59 minutes booked against 6816. Foreman E C Mills examined the loco which was pretty much bunged up and also had a defective brick arch. He passed it on to Horton Road for attention there.

No names, no packdrill, but Barnwood shed had its suspicions about the actions of the crew following the failure of 45464 on 30 October, working the previous evening's 4M32, 10.0pm Avonmouth-Water Orton banana special. This was delayed by 109 minutes at Eckington after 45464 got a hot box on the tender and was replaced by Barnwood's 48463. The Barnwood report stated 'Upon examination at this depot the axlebox pad was found to be completely destroyed although there was a good supply of new oil in the box. This appears to suggest that the oil was fed to the box after the overheating took place.' 45464 was still out of traffic at Barnwood on 20 November.

Barnwood Driver A Moxey and Fireman Rickards had problems on two consecutive turns along the Honeybourne line. On 31 October, they were crewing the 3.25am Rogerstone-Oxford with 48611 of Woodford Halse depot. Upon arrival at Honeybourne, the driver noticed that the lead plug was weeping, so the engine was put on Honeybourne loco and the fire dropped by the relieving Worcester man. Driver Moxey wrote 'At no time were we short of water and both gauge frames were working correctly.'

After that episode, the next night they had ex-Crosti 9F 92028 on the 10.20pm Cardiff-Woodford Halse. This was short of steam and they lost 25 minutes on the journey, due to the elements blowing. The driver added 'I understand this engine has been in this condition for a long while.' Perhaps not one of Saltley's finest!

A GWR 2-8-0 replaced an LMS 2-8-0 on 13 November on the 12.30pm Helpston (between Peterborough and Stamford) -Stoke Gifford freight. 48095 got a hot box on the tender, causing a two hour delay until 3851 took over. 48095 was laid-up at Gloucester, while parts were sent to Crewe Works for reconditioning.

On Saturday/Sunday 7/8 December there were three failures in quick succession on freights. Driver F York and Fireman R Gay

of Barnwood had the 8.15pm from Washwood Heath to Westerleigh, 8V28, with 44180. The driver complained he was unable to see the signals because of steam blows at the front of the engine, which also affected maintaining the boiler pressure. So 44840 replaced 44180 at Gloucester. 44180 had recorded 74,800 miles since a heavy intermediate overhaul at Derby dated 6 April 1960, which sounds more plausible than 44263's mileage referred to earlier.

Exactly the same problem of steam blows at the front end afflicted another Saltley 4F 44571 which was turned out for the 11.30pm Saturdays only Gloucester Docks Branch-Washwood Heath, 8M54. It returned to Barnwood shed and was then sent light engine back to its home depot for attention. The trouble was caused by the cylinder cocks sticking in the open position.

Meanwhile Barnwood Driver G Noakes and Fireman A Green were making their way home with 48089 – of Heaton Mersey shed – on the 9.40pm Washwood Heath-Westerleigh, 8V27. But after hearing an unusual noise, they stopped at Bredon and discovered the left driving wheel spring had broken. They worked the train to Cheltenham High Street where it was recessed and carried on light engine to Barnwood.

The 9.40pm was also in trouble on 9 December, again at Bredon. Another loco, 45184, from a relatively rare shed – Chester – gave up when the back set of firebars collapsed into the ashpan. This time, a fresh engine, 92083, took over there with an hour's delay.

Three engines which failed on 13/14/15 December were returned light to their home depots for attention, as Barnwood claimed it already had a heavy repair burden. All were '9F's, Nos. 92029 and 92155 from Saltley and 92000 of Bristol Barrow Road. The latter was on an Avonmouth-Water Orton train, which went north behind newly-repaired 48095 on 14 December. 92155's replacement on the 1.0am Washwood Heath-Westerleigh class D freight on 15 December was former Barrow Road star 'Jubilee' 45660 *Rooke*, now shedded at Shrewsbury.

45685 *Barfleur* – one of the three 'Jubilees' still allocated to Barrow Road, the others being Nos. 45682 and 45690 – was short of steam on 17 December with the 8.40am Bristol-Sheffield express, 1E64, causing a half-hour delay and it retired at Gloucester in favour of D32. As mentioned earlier, this particular train was a favourite for steam haulage and, on 7 December, Burton depot's 45598 *Basutoland* also failed at Gloucester short of steam on the working.

Another northbound express, 1N84, 12.55pm Bristol-Bradford, saw steam on occasion, one such being Sunday 22 December, when 45561 *Saskatchewan* failed short of steam at Gloucester. Barnwood was able to provide classmate 45674 *Duncan* which took over at Engine Shed Junction, just 16 minutes down.

A notable locomotive the previous day on the 11.50am York-Bristol, 1V43, was 'Britannia' 70007 Coeur-de-Lion instead of the usual diesel, 101 minutes late at Cheltenham.

By the end of 1963 diesels were in the ascendancy but regularly suffering failures. Train 1M77, the 12.20pm Cardiff-Derby, was normally a DMU, but to provide extra passenger capacity over the

Above: Woodford Halse shed was renowned for its large allocation of WD 2-8-0s, but Stanier 8F 48611 was also stationed there. In this picture it is at Cheltenham Lansdown Junction having travelled down the Honeybourne line on a freight bound for South Wales. On 31 October 1963 a leaking lead plug caused the loco's failure at Honeybourne on a Rogerstone-Oxford freight, with Barnwood men on the footplate.   *A W V Mace collection*

holiday period, it was engine and coaches on 28 December with diesel-hydraulic D7030. This became a locomotive casualty at Cheltenham Lansdown. 45369 was commandeered from a down working and towed the train, tender-first, to High Street where both locos were replaced by 44981, which also did the return working, 1V72, 5.20pm Derby-Cardiff, 32 minutes late. The next day saw more diesel woe, with 1V43 running 103 minutes late at Cheltenham with D60 and 'Black Five' 44888 coupled inside, probably for steam heating.

Until the advent of Brush Type 4s on the Birmingham-Bristol line in 1964, there were not enough diesels to cover every failure and even some prestigious workings were still occasionally steam – a couple of examples were the down 'Devonian' on 18 January 1964 which had 45442 around 30 minutes late at Cheltenham; and the 7.25pm Bristol-Newcastle mail on 2 March 1964 when 44814 replaced a failed diesel at Berkeley Road. It was the first recorded instance of steam on this train since 7 April 1963 – but it was steam again less than a month later, on 3 April 1964, with 45564 New South Wales, 31 minutes behind schedule at Cheltenham, at the head of 12 coaches, including a sleeping car.

And finally...

Barnwood was an old-fashioned steam depot which had no part to play in the rapidly modernising British Railways and it closed officially in May 1964, with the remaining staff and locos transferred to Gloucester Horton Road shed. For a while this made the latter rather crowded and some engines were sent to lay over at Barnwood – for example, a visit on 14 June 1964 found: 42904; 44137; 45221; 45369; 48368; 70015; 73094; 78009; 92112. (Horton Road shed hosted 61 steam locos on the same date.) Barnwood was also used

for some months after closure for staging withdrawn locos on their way to scrapyards.

Pat Cook recalls an interesting incident one day when he was on duty at Gloucester Control. A diesel on a down afternoon express failed at Coaley Junction. The nearest available locomotive was a 9F on a down freight at Stonehouse. Pat managed to contact the signalman at Stonehouse just as the 9F was about to be given the right away to proceed. It was arranged for the 9F to go light engine to Coaley and push the express to Charfield. At the latter, the 9F ran along the down loop, coupled to the front of the train, and headed it, complete with failed diesel, to Bristol. This left the freight at Stonehouse without motive power. A quick call to Westerleigh marshalling yard found a loco there which was able to travel up light and take the train away – one of the S&D 7F 2-8-0s, thus providing a treat for anyone that happened to be observing at the lineside.

On 29 April 1964 Derek Smith was the Barnwood fireman on Saltley 8F 48351 which was requisitioned from a Bath Gasworks coal train at Standish Junction to assist the 7.35am Nottingham-Bristol passenger, 1V31, after the Sulzer diesel hauling it failed at Stonehouse. 48351 ran wrong line to Stonehouse to rescue the failure. Luckily 48351 was one of the class modified with balanced wheels, so did not ride too badly, achieving a top speed of 46 miles per hour on the journey! Ironically, when the train arrived at Temple Meads, the diesel driver tried to start his engine – and it burst into life. But the diesel's vacuum was still not working, so 48351 towed it to the shed after arrival.

With a real mix of engines available at Horton Road after Barnwood's closure, some interesting workings occurred. On 2 June 1964, the 'Cheltenham Spa Express', loaded to nine coaches, was hauled from Gloucester to Cheltenham by former Barnwood resident 43887 – deputising for the usual GW 2-6-2T, which had presumably failed – probably the only time this train had Midland 4F haulage... unless anyone knows differently!

OTHER CASUALTY REPORTS in 1963

Date	Loco	Home shed	Train	Problem	Place	Delay (mins)	Notes
09/01/63	44092	21A	12.5am WWH - Bristol freight	MECH	GLOS	0	1
17/01/63	44131	21A	8.55pm Avonmouth-Water Orton class D	VBI	GLOS	18	2
23/01/63	44858	21A	4.5pm Newcastle - Bristol class A	SOS	ER		3
25/03/63	44179	21A	8.55pm Avonmouth - Water Orton class D	SOS	ER	20	
30/10/63	44605	21A	6.25pm Stoke Gifford - WWH freight	SOS		97	4
07/11/63	92136	21A	2.0pm WWH - Stoke Gifford freight	SOS	ER	82	5
12/11/63	45712	17A	6.45am Gloucester- Sheffield class B	SOS			6
24/11/63	44160	21A	7.25pm WWH-Westerleigh freight	HOT	ER	57	7

Notes

1 Steam brake exhaust pipe broke, repaired at Barnwood. Loco worked forward
2 44180 took train north. 44131 repaired at Barnwood
3 Fresh loco from Gloucester
4 Loco failed at Berkeley Road. 73028 took train. 44605 light to Saltley for attention
5 Loco worked by Tyseley men, had done trip work prior to this working, dirty fire blamed
6 Time lost at Ashchurch and Worcester from where 44945 provided assistance
7 48133 took train from Gloucester. 44160's wheels and boxes sent away for repair

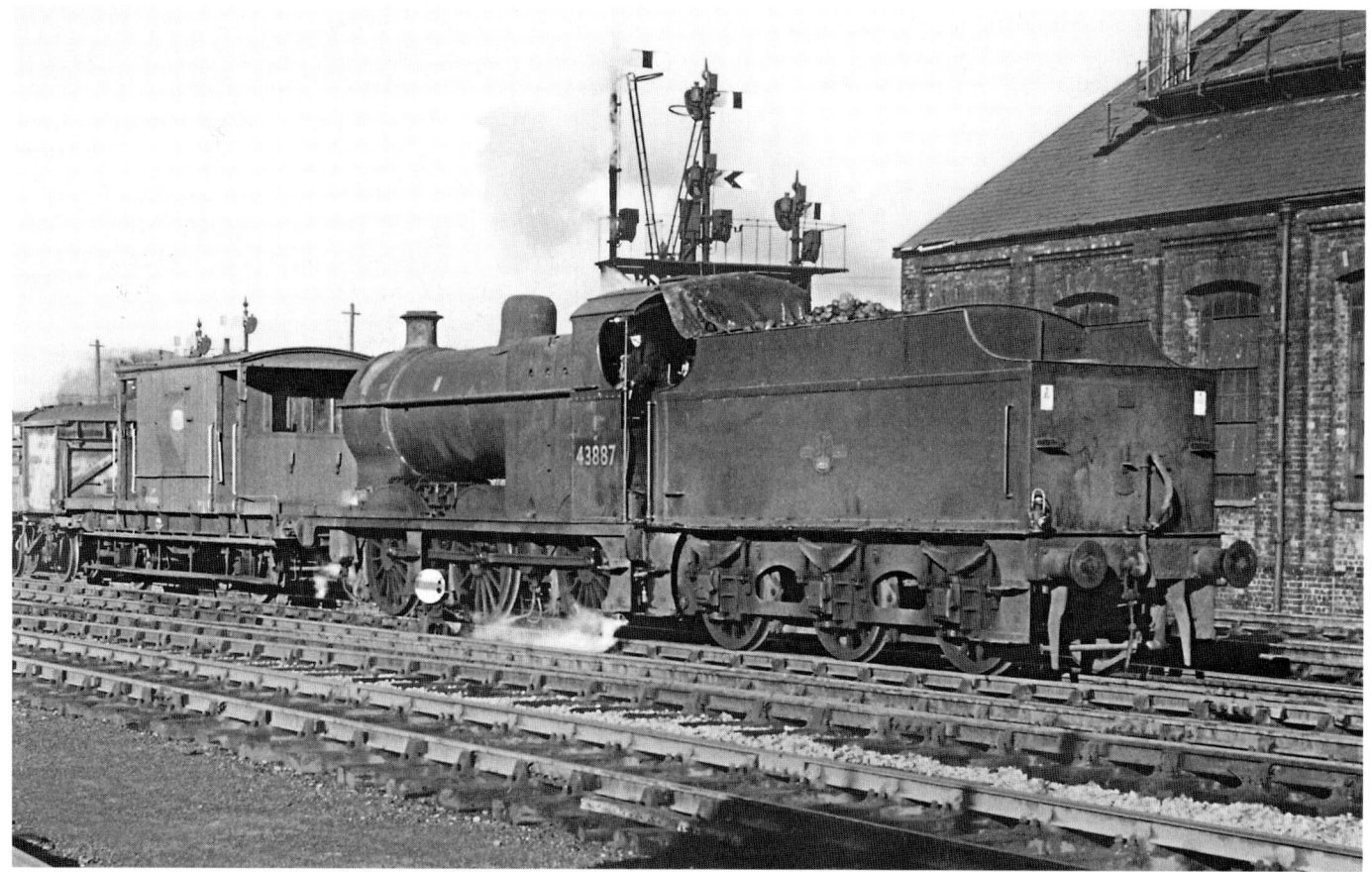

Above: Former Barnwood resident 4F 0-6-0 43887 does some shunting alongside its then current home, Gloucester Horton Road, in January 1965 shortly before withdrawal from service. Possibly one of the loco's finer moments occurred on 2 June 1964 when it was an unusual choice to haul the 'Cheltenham Spa Express' from Gloucester Central to Cheltenham St James. One suspects it was a last-minute replacement for the normal GW Prairie tank.

CHAPTER SIX

BARNWOOD LOCOMOTIVE DEPOT

Above: Barnwood shed yard would usually contain an engine or two bereft of a wheel set and in June 1959 it is 4F 44553 of Bristol Barrow Road and at least one other of the class. Behind those two is 2P 4-4-0 40489.
Photomatic

Whilst rationalisation of facilities duplicated by the 'Big Four' railway companies might have been anticipated, not much had altered outwardly at Gloucester in the decade following nationalisation. The ex-LMS station at Eastgate had its counterpart in the ex-GWR station at Central and passengers transferring between the two were faced with a long walk across the connecting footbridge. Gloucester Docks still had two virtually separate railways, shunted by Midland engines – Deeley 0-4-0T – on one side and Western engines – 0-6-0PT – on the other. The various Western and Midland goods yards were still in business. And there were two engine sheds – the former Midland roundhouse at Barnwood, opened in July 1894, with the GWR's straight sheds at Horton Road. Loco staff were emphatically either 'Midland men' or 'Western men' – and never the twain shall meet. But organisational changes meant the ex-Midland/LMS Birmingham – Bristol main line and branches were transferred from the London Midland Region to the Western Region from 2 April 1950 for certain administrative purposes, with the 'border' being at Barnt Green. However it was February 1958 before the Western Region took over completely and Barnwood's shed code was altered from the LMR's 22B to the WR's 85E (changed again to 85C from January 1961); Bristol Barrow Road altered from 22A to 82E, and Bath Green Park, which had been 22C at the start

of the BR era and then 71G, later became 82F.

Fred Cole started his stint as Shed Master in March 1958 and recalls that, at that time, he reported to Derby on locomotive matters and Euston on staff matters, despite Barnwood being in the Western Region; while the District Office was situated at Worcester where the Locomotive Superintendent was Bill Sidwell. Fred's predecessor at Barnwood, Sam Knowles, moved to Horton Road shed. As related in the Introduction, this changed in 1960 and again during 1963 when Barnwood came under the Divisional Office at Bristol.

The location of Gloucester Barnwood motive power depot was shown as follows in that indispensable guide *The British Locomotive Shed Directory*: 'The shed is on the north side of the main Gloucester-Cheltenham line about a half-mile north of the junction at the end of the two Gloucester stations. (The line is running east – west by the shed.) The yard is visible from the line. Walking time from either Gloucester Eastgate or Gloucester Central stations – 15 minutes.'

In Midland Railway days, Barnwood was provided with shear-legs, located inside the roundhouse, and axle box boring machinery. It had an 'outstation' shop where heavy repairs were undertaken, though this was reduced in status in LMS days to become a fitting shop for the running shed. The LMS invested in a wheel drop, located in the yard, with a basic tin shelter covering it, and a new turntable in the 1930s, but the shed's coal stage was not modernised as happened with some depots such as Bristol Barrow Road.

Barnwood was not a glamorous 'top shed' with gleaming 'Pacifics' but it did what needed doing on the allocated jobs with its own modest stable of locomotives of mainly Midland Railway design, and 'foreign' engines. The majority of the depot's work was centred on the former Midland Railway Birmingham – Bristol/Bath main line and associated branches. The shed had passenger links covering express and stopping trains on the Derby – Birmingham – Bristol and Bath line. The freight links work was also mainly on the same route and associated branches. There were turns on the ex-GWR Honeybourne line, mostly with freights which used to travel via Ashchurch, Broom Junction and the Stratford-on-Avon & Midland Junction line to Woodford Halse before a new curve constructed at Stratford caused these workings to be shifted to the Honeybourne route in 1960.

Other depots also provided engines and crews for workings through Gloucester, prominent amongst them being Saltley, Bristol Barrow Road, Bath Green Park, Derby, Sheffield Millhouses and Leeds Holbeck, adding variety to train observations.

By the start of this survey in 1958 a number of older locomotive types had been replaced – Barnwood no longer housed Johnson 0-4-4T and 0-6-0T, while 4-4-0s, once so common, were nearing the end of their working lives. Horton Road's 4-4-0s, 'Dean Goods' 0-6-0s, 'Star' and 'Saint' 4-6-0s and the older types of 0-6-0PT had given way to more recent GW types in the early 1950s.

34 locos were shedded at Barnwood in February 1958, all ex Midland Railway or LMS types:
40489 / 40540 / 40933 / 41049 / 41093 / 41095 / 41123 / 41535 / 41537 / 41900 / 43337 / 43373 / 43520 / 43645 / 43754 / 43853 / 43887 / 43924 / 44035 / 44045 / 44123 / 44167 / 44209 / 44264 / 44272 / 44296 / 44567 / 46401 / 47417 / 47422 / 47506 / 47539 / 47623 / 58165
A couple of Barnwood locos were sub-shedded at Tewkesbury for trains between Ashchurch and Upton-on-Severn; shunting at Ashchurch; freight from Ashchurch to Evesham and Ashchurch to Cleeve. The most modern loco on the roster, Ivatt 2-6-0 46401, replaced ancient Johnson 0-4-4T 58071 at Tewkesbury in 1956 and shared the duties for a while with Stanier 0-4-4T 41900. Engines which worked from Birmingham on trains via Redditch and Evesham to Ashchurch also utilised the facilities at Tewkesbury shed.

Another branch under Barnwood's auspices, Coaley Junction to Cam and Dursley, was normally worked by a WR 0-6-0PT, often 1605, which stabled at Dursley but it does not show up in the roster, being officially allocated to 85B Horton Road. 1605 took over from ex-Midland Railway 0-6-0Ts which used to work the branch, the last ones were 41720 and 41748.

Barnwood also provided an engine, usually an 0-6-0T, to shunt the former Midland goods yard at Cheltenham High Street and this stabled at Cheltenham Malvern Road loco depot. After the Western Region takeover in 1958, Barnwood received a couple of BR Standard Class 4 4-6-0s in August – 75009 and

75023 – as a start towards replacing the 4-4-0 fleet. 75009 was no stranger to the area, having spent a few years allocated to Cardiff Canton – during which time it worked stopping trains from Cardiff to Gloucester and also sometimes the heavily-loaded Newcastle express. Two ex-GWR pannier tanks – 7723 and 7756 – arrived in November 1959 to work from Tewkesbury shed, making 41900 redundant; it went into store, but found further use at Leamington Spa in the summer of 1960. 46401 had by then moved on. Gloucester's last Compound – 41123 – was withdrawn in December 1959, but remained at the depot until early May 1960 before going for scrap. The 2Ps – 40489 and 40540 – hung on for a while longer, the latter engine not being officially withdrawn until 1962 though by then it had been in store at Barnwood for some time.

Barnwood's allocation at the end of 1959 was 31 locos:
7723/ 7756/ 40489/ 40540/ 41535/ 41537/ 41900/ 43337/ 43645/ 43754/ 43853/ 43887/ 43924/ 44035/ 44045/ 44123/ 44167/ 44209/ 44264/ 44272/ 44296/ 44567/ 47417/ 47422/ 47506/ 47539/ 47623/ 75009/ 75023/ 90565/ 90685.
The two WD 2-8-0s did not stay long, being transferred away in the summer of 1960. Still not shown in the allocation at this date was 0-6-0PT 1605.

Some weekday turns worked by Barnwood locos, 1960/61:

Passenger
    2.3pm Saturdays excepted Birmingham New St-Worcester
       stopper
    5.15pm Bristol-Birmingham stopper
    4.5pm Gloucester-Bristol stopper
    Gloucester Eastgate station pilot
    Ashchurch-Tewkesbury-Upton-on-Severn
    Coaley Junction-Dursley

Freight and Shunting
    4.42am class 'K' Gloucester-Bromsgrove
    8.55pm class 'H' Gloucester-Bristol
    9.45pm class 'D' Gloucester-Water Orton
    Ashchurch-Tewkesbury-Upton-on-Severn
    Ashchurch-Evesham
    Stroud and Nailsworth branch freights
    Sharpness and Dursley branches
    Trip workings to Hempstead and Quedgeley
    Shunting Gloucester Docks (41535 and 41537)
    Shunting and trip work, Gloucester and Cheltenham
    'Ballast engine' used for Civil Engineer's duties (Midland
       2F 58165 at the start of this survey, later a 3F)

Some turns worked by Barnwood crews, 1960/61:

Passenger Weekdays unless otherwise shown
    9.2pm Bradford-Bristol, worked from Derby to Gloucester
    11.45pm Bristol-Derby parcels, worked from Bristol
       to Birmingham
    1.10am Bristol-Sheffield mails, worked from Gloucester
       to Birmingham
    7.5pm Newcastle-Bristol mails, worked from Birmingham
       to Gloucester

Above: Barnwood's last Compound, 41123, was a casualty of modernisation, being withdrawn in December 1959, though quite active until the end. It remained at the back of the roundhouse siding for some months until towed away for scrap in May 1960. This picture is dated 24 April 1960 and 41123 still looks very respectable with front number and works plate intact, and lamp on the front.   *Brian Miller collection*

5.20am Mondays excepted Birmingham-Worcester fish train

7.0am Birmingham-Bristol stopper, worked to Gloucester

6.18am Derby-Bristol, worked from Gloucester to Bristol

8.30am Cardiff-Newcastle, worked from Gloucester to Birmingham

7.35am Nottingham-Bristol, worked from Gloucester to Bristol

10.25am Manchester-Bournemouth 'The Pines Express', worked from Birmingham to Gloucester. The duty at Birmingham included transferring the Sheffield portion of the train onto the main Manchester train, with the engine of the former then working from New Street to Bath.

1.45pm Gloucester-Worcester-Birmingham stopper (change engines at Worcester)

1.32pm Derby-Bristol parcels, worked from Gloucester to Bristol

6.30pm Bristol-Birmingham stopper, worked to Gloucester

7.25pm Bristol-Newcastle mails, worked from Gloucester to Birmingham

Station pilot at Gloucester Eastgate

1.20pm Sundays Bristol-York, worked from Gloucester to Birmingham

12.15pm Sundays York-Bristol, worked from Birmingham to Gloucester

8.15pm Sundays Gloucester-Catterick Camp, worked to Derby

9.5pm Sundays Bradford-Bristol, worked from Derby to Gloucester

Freight – Weekdays

10.10pm class 'J' Washwood Heath-Westerleigh, worked from Gloucester

2.45am class 'C' Washwood Heath-Westerleigh, worked to Gloucester

4.42am class 'K' Gloucester-Bromsgrove

4.0am class 'F' Bath-Lawley Street, worked from Gloucester

7.15am class 'H' Honeybourne-Severn Tunnel Junction, worked from Honeybourne to Gloucester

2.55am class 'F' Cardiff-Woodford Halse, worked from Gloucester to Honeybourne

1.0pm class 'D' Washwood Heath-Gloucester

11.0am class 'D' Avonmouth-Bromford Bridge tanks, worked from Gloucester

11.20am class 'H' Bristol-Washwood Heath

2.0pm class 'C' Evesham-Water Orton

2.55pm class 'C' Bristol-York, worked to Gloucester

3.25pm class 'J' Ashchurch-Evesham

4.20pm class 'C' Burton-on-Trent-Bristol 'beer', worked from Birmingham Camp Hill to Gloucester

12.36pm Toton-Westerleigh, worked from Bromsgrove to Gloucester

7.15pm class 'H' Gloucester-Washwood Heath

7.0pm class 'D' Bristol-Derby, worked from Gloucester to Birmingham

9.45pm class 'D' Gloucester-Water Orton

2.25pm class 'D' Fawley-Bromford Bridge Esso tanks, worked from Gloucester to Bromford Bridge

Various branch freights, trip work and shunting.

2.32pm class 'D' Gloucester-Lawley Street ('The Snob').

SOME ENGINES REPAIRED AT BARNWOOD during 1960 and 1961 (per BR forms 9214)

Loco	Shed	Date A	Date B	Date C	Repair	Parts	Notes
7201	86G	26-7-60		17-8 60	HB	SW	
47506	85E	27-7-60		20-8-60	HB	DE	
44076	2E	28-7-60		25-8-60	HB	DE	
7756	85E	17-6-60		28-8-60	HB / 6ex	SW	
45597	55A	2-8-60		2-9-60	HB	CR	
73090	84G	25-8-60		12-9-60	HB	SW	1
3828	84J	1-9-60		25-9-60	HB	SW	2
44213	21A	19-9-60		5-10-60	HB	DE	
44045	85E	27-9-60		12-10-60			3
44296	85E	6-10-60		19-10-60	HB	BR	
43855	21A	7-10-60		25-10-60	HB	DE	4
73136	17A	23-9-60		27-10-60		DE	5
90544	86A	16-9-60		28-10-60	HB	CR	
7788	85E	20-10-60		4-11-60		SW	6
45263	21A	27-9-60		11-11-60	HB	DE	
42764	21A	13-10-60		26-11-60	HB	HO	
43812	21A	16-12-60		11-1-61	HB	BR	
44943	55A	9-12-60		16-1-61	HB	DE	
47308	85D	3-1-61		19-1-61	HB	SW	
2890	82C	4-1-61		30-1-61	HB	SW	
44211	21A	5-1-61	19-1-61	7-2-61	HB	DE	
43754	85C	14-1-61	23-1-61	8-2-61	HB	DE	
43507	21A	6-2-61	6-2-61	25-2-61	HB	DE	
45668	17A	31-1-61	31-1-61	27-2-61	HB	DE	
44551	17B	20-2-61	21-2-61	6-3-61	HB	DE	
6879	84E	15-2-61	16-2-61	10-3-61	HB	SW	
41537	85C	27-2-61	27-2-61	13-3-61	HB	DE	
44962	21A	4-2-61	8-2-61	14-3-61		DE	7
47308	85D	1-3-61	1-3-61	14-3-61	6ex		
92241	88A	25-2-61	2-3-61	24-3-61	HB	CR	
43963	21A	11-3-61	18-3-61	30-3-61	HB	SW	
4961	81D	10-3-61	14-3-61	6-4-61	HB	SW	8
75009	85C	30-3-61	30-3 61	14-4-61		SW	9
42421	21A	30-3-61	54-61	21-4-61	HB	DE	
43938	21A	19-4-61	19-4-61	2-5-61	HB	DE	
42827	21A	13-4-61	14-4-61	10-5-61	HB	HO	
44659	21A	244-61	24-4-61	11-5-61	HB	DE	
3860	88A	14-4-61	24-4-61	16-5-61	HB	SW	
44171	21A	1-5-61	2-5-61	20-5-61	HB	DE	
44102	82G	13-5-61	16-5-61	28-5-61	HB	DE	
44482	41E	13-5-61	24-5-61	4-6-61	HB	DE	
44184	21A	29-5-61	5-6-61	21-6-61	HB	DE	
42897	17B	16-5-61	31-5-61	22-6-61	HB	DE	
43507	21A	9-6-61		26-6-61	HB	DE	
43680	21A	25-5-61	2-6-61	28-6-61	HB	DE	
44296	85C	8-6-61	8-6-61	1-7-61	6ex		
43853	85C	10-6-61	10-6-61	1-7-61	HB	DE	
45668	17A	2-6-61	6-6-61	16-7-61	HB	DE	
44851	21A	7-6-61		31-7-61	HB	CR	
45097	24L	24-8-61			HB		
44586	55B	15-7-61	17-7-61	9-8-61	HB	DE	
45659	55A	19-7-61	19-7-61	9-8-61	HB	DE	
44580	21A	2-8-61	2-8-61	14-8-61	HB	DE	
92216	88A	20-7-61	22-7-61	18-8-61	HB	SW	
92213	71A	27-7-61	1-8-61	21-8-61	HB	CR	
45269	21A	1-7-61	12-7-61	22-8-61	HB	CR	

SOME ENGINES REPAIRED AT BARNWOOD during 1960 and 1961 (per BR forms 9214) continued

Loco	Shed	Date A	Date B	Date C	Repair	Parts	Notes
46122	21A	19-7-61	20-7 61	23-8-61	HB	CR	
73021	87F	11-8-61	12-8-61	28-8-61	HB	SW	
44045	85C	3-8-61	14-8-61	28-8-61	HB	DE	
6848	86G	17-8-61	23-8-61	16-9-61	HB	SW	
44580	21A	26-8-61	2-9-61	17-9-61	HB	DE	
44272	85C	24-8-61	24-8-61	18-9-61		DE	10
92242	86A	19-8-61	29-8-61	3-10-61	HB	CR	
3801	86E	17-9-61	19-9-61	15-10-61	HB	SW	
47539	85C	19-9-61	2-10-61	19-10-61	HB	DE	
90492	40E	22-9-61	22-9-61	25-10 61	HB	CR	
9740	82C	16-9-61	18-9-61	28-10-61	HB	SW	11
73021	87F	29-8-61	31-8-61	30-10-61	HB	SW	
45447	21A	4-10-61	4-10-61	31-10-61		DE	12
8401	85D	6-10-61	9-10-61	31-10-61	HB	SW	
5049	82B	6-10-61	7-11-61		HB	SW	13
8725	82E	27-9-61	19-10-61	10-11-61	HB	SW	
43645	85C	27-10-61	13-11-61			DE	14
47623	85C	2-11-61	2-11-61	27-11-61	HB/6ex	DE	
6807	85A	8-11-61	8-11-61	2-12-61	HB	SW	
5640	82B	11-11-61	12-11-61	17-12-61	HB	SW	

Notes

1	Grease escaping from Timken roller bearing axleboxes
2	Connecting and coupling rods also sent to Swindon
3	15 broken stays changed
4	New lubricator fitted
5	Right-hand connecting and coupling rods, right hand cylinder front cover and piston renewed
6	Ejector box repair
7	Brake gear defective, combination iever and union link broken
8	Failed at Worcester on 7-3-61 on 9.15am Paddington-Hereford passenger
9	Right valve crosshead loose
10	Defective regulator valve
11	Failed at Swindon on 29-8-61 on 5.55pm Swindon-Chisledon passenger
12	Left-hand cylinder front cover broken; replaced
13	Failed at Swindon on 1-10-61 on 11.40pm Paddington-Bristol passenger
14	Pin seized in left-hand die block

Key:

DATE A: Date out of service; DATE B: Date to Works: DATE C: Date into traffic; HB: Hot box; 6ex: No.6 examination; SW: Parts from Swindon Works; DE: Parts from Derby Works; CR: Parts from Crewe Works; HO: Parts from Horwich Works; BR: Parts from Bristol Barrow Road.

Left: A welcome sight for the running foreman was a 'No Repairs' ticket put in by a driver when an engine came to shed. This one is from Toton based 8F 48363 which was at Barnwood on 16 October 1963. Other tickets show that 48363 was also at Toton that day, while it was at Bescot the previous day and Saltley on 17 October. Tickets from the sheds visited by the loco were forwarded to its home depot.

Above: 9F 92241 was a Cardiff Canton loco which ended up at Barnwood for attention in February 1961. Parts were sent to Crewe for repair and the loco was out of traffic for a month.   *Kidderminster Railway Museum*

Below: 0-6-0T 47308 of Bromsgrove sits in the yard at Barnwood awaiting attention to a hot box in January 1961. Bromsgrove's shed code is crudely painted on the smoke box, it had just changed from 85F to 85D. After this sojourn, 47306 was back in March for a couple of weeks, undergoing a valves and piston examination.

LOCO MILEAGES SHOWN IN CASUALTY REPORTS

Loco	Home shed	Failure date	Mileage	Last works overhaul Place	Date	Type
73031	82E	7.11.59	78118	DERBY	23.12.57	GENERAL
44296	85E	22.12.59	700	DERBY	24.11.59	
90685	85E	26.12.59	33000	CREWE	25.6.58	
43242	21A	9.1.60	22262	DERBY	2.7.58	GENERAL
45683	41C	13.1.60	52500	CREWE	24.12.58	LI
42823	21A	19.1.60	23380	HORWICH	5.6.59	INT
44520	21A	1.2.60	11200	DERBY	1.2.59	
48220	21A	3.2.60		HORWICH	19.6.58	INT
42761	21A	13.2.60	64800	HORWICH	28.8.57	INT
42900	21A	16.2.60	34000	HORWICH	27.2.59	GENERAL
43047	21A	22.2.60	67000	HORWICH	16.10.57	INT
42825	17B	23.2.60	61914	HORWICH	3.10.57	
48672	18A	24.2.60	35000	DERBY	28.9.58	GENERAL
43940	21A	24.2.60	19200	DERBY	16.1.59	GENERAL
43975	21A	26.2.60	29300	DERBY	17.11.58	GENERAL
44851	17A	5.3.60	50264	CREWE	18.10.58	GENERAL
42790	21A	14.3.60	75000	HORWICH	30.1.58	
75004	82E	17.3.60	14700	SWINDON	23.9.59	H GEN
44662	55A	20.3.60	66500	CREWE	25.9.58	GENERAL
90565	85E	25.3.60	11500	CREWE	7.10.59	
73144	17C	21.4.60	150000	DERBY	28.12.56	NEW
44187	21A	26.4.60	15402	DERBY	5.10.59	H INT
43359	21A	12.5.60	86291	DERBY	23.8.55	
45669	1A	14.5.60	68006	CREWE	13.11.58	
42859	1A	17.5.60	87000	HORWICH	28.4.59	H INT
45579	17A	19.5.60	58000	CREWE	29.4.58	GENERAL
45610	17A	26.5.60	2000	CREWE	12.5.60	INT
44755	55A	26.5.60	91917	CREWE	30.5.58	INT
45130	5B	4.6.60	66452	CREWE	9.10.58	
45696	3D	17.6.60	69574	CREWE	19.9.58	
44002	41D	18.6.60	17581	DERBY	10.8.59	CASUAL
75002	85E	21.6.60	76210	SWINDON	28.11.57	GENERAL
44337	55E	23.6.60	67000	DERBY	26.4.57	
45519	82E	12.7.60	14000	CREWE	12.3.60	LI
45076	26A	19.7.60	39911	CREWE	30.10.59	
73000	18B	22.7.60	60002	DONCASTER	22.1.59	
45597	55A	1.8.60	91500	CREWE	18.4.59	INT
92070	21A	3.8.60	67132	CREWE	24.11.58	
42764	21A	5.8.60	28583	HORWICH	30.10.59	
43924	85E	6.8.60	36138	DERBY	16.2.59	GENERAL
43853	85E	10.8.60		DERBY	10.1.58	
44179	21A	21.8.60	37032	DERBY	25.3.59	
44167	85E	22.8.60	42588	DERBY	22.1.59	
43951	21A	22.8.60	70000	DERBY	1.5.57	
75009	85E	22.8.60	41523	WOLVERHAMPTON	27.6.59	INT
44296	85E	5.10.60	19826	DERBY	24.11.59	H INT
44209	85E	7.10.60	13100	DERBY	3.5.60	H INT
42846	21A	11.11.60	59309	HORWICH		
43521	21A	24.11.60	3900	DERBY	11.8.60	H INT
48669	21A	1.12.60	33800	DERBY	23.11.59	GENERAL
44226	21A	7.12.60	36509	BOW	10.7.59	INT
45607	41C	8.12.60	14500	CREWE	23.9.60	LI
43924	85E	16.12.60	43103	DERBY	16.2.59	GENERAL
44272	85E	20.12.60	354	DERBY	9.11.60	LI
44211	21A	4.1.61	36500	DERBY	11.6.59	GENERAL
44818	21A	7.1.61		CREWE	28.2.59	H GEN
45418	3D	7.1.61	44334	CREWE	5.1.60	
45607	41C	18.1.61	29500	CREWE	23.9.60	LI

73136	17A	26.1.61	54000	DERBY	27.5.59	
45577	82E	30.1.61	22000	CREWE	3.9.60	LI
44962	21A	3.2.61	50000	CREWE	31.12.59	
43507	21A	4.2.61	75650	DERBY	29.1.54	GENERAL
92007	82E	6.2.61	31248	SWINDON	XX.1.59	H INT
92248	82E	7.2.61	52000	CREWE	2.12.58	NEW
45602	41C	14.3.61	22500	CREWE	4.11.60	GENERAL
45506	82E	20.3.61	4000	CREWE	21.1.61	L CAS
45654	41C	3.4.61	45500	CREWE	26.5.60	INT
73003	82E	4.4.61	63711	SWINDON	12.11.59	H INT
45564	55A	4.4.61	81260	CREWE	26.11.59	INT
45690	82E	5.4.61	44712	CREWE	6.5.60	H GEN
92205	71A	5.4.61	61000	SWINDON	XX.4.59	NEW
45447	21A	10.4.61	16500	CREWE	13.12.60	
92007	82E	10.4.61	31605	SWINDON	XX.10.59	
42827	21A	12.4.61	51500	HORWICH	14.10.59	
45569	55A	19.4.61	76458	CREWE	29.1 2.59	INT
44171	21A	21.4.61	48554	DERBY	27.2.59	
75009	85C	21.4.61	56323	SWINDON	27.6.59	
44659	21A	24.4.61	49769	CREWE		
44857	55A	28.4.61	102059	CREWE	25.3.59	INT
45651	82E	28.4.61	26000	CREWE	10.11.60	H GEN
42756	17B	2.5.61	43398	HORWICH	24.6.59	GENERAL
44102	82E	12.5.61		DERBY	28.4.61	GENERAL
45612	17A	16.5.61	5000	CREWE	29.3.61	INT
44333	21A	17.5.61	39000	DERBY	18.10.59	H INT
73043	41D	29.5.61	106700	DONCASTER	5.7.58	INT
45557	17A	6.6.61	48515	CREWE	24.4.61	GENERAL
44436	17B	14.6.61	58977	DERBY	29.4.58	H GEN
45590	41C	14.6.61	43479	CREWE		LI
1605	85C	23.6.61	75250	SWINDON	24.4.56	H GEN
45097	24A	23.6.61	51829	CREWE	26.3.59	
44858	21A	25.6.61	26750	CREWE	8.12.60	H INT
48669	21A	1.7.61	50250	DERBY	24.11.59	H GEN
44818	21A	4.7.61	99000	CREWE	28.2.59	H GEN
44013	21A	12.7.61	16500	DERBY	6.10.60	H INT
44296	85C	27.8.61	33327	DERBY	24.11.59	H INT
46103	21A	18.10.61	36000	CREWE	24.1.61	
42823	21A	19.10.61	9500	HORWICH	27.6.61	
43645	85C	27.10.61	509	DERBY	21.9.61	H INT
46137	21A	7.11.61	25500	CREWE	11.1.61	H INT
92139	21A	13.11.61	149000	CREWE	NEW	
92156	18A	20.11.61	103000	CREWE	15.11.57	NEW
44841	21A	22.11.61	83000	CREWE	14.1.60	H GEN
45658	55A	26.11.61	14000	CREWE	22.7.61	INT
47417	85C	20.12.61	55052	DERBY	11.4.58	GENERAL
73096	85C	28.1.63	53400	DONCASTER	24.3.61	GENERAL
73094	85C	26.2.63	62757	DONCASTER	3.11.60	INT
45569	55A	10.3.63	53000	CREWE	18.1.62	GENERAL
44092	21A	23.5.63	22500	DERBY	29.3 63	INT
92157	21A	23.5.63	157700	CREWE	20.11.57	NEW
48420	85C	12.6.63	49000	CREWE	28.8.61	H INT
90439	36C	28.6.63	27000	GORTON	XX.5.62	
92152	21A	24.9.63	7000	CREWE		CASUAL
48463	85C	25.9.63	47337	CREWE	15.11.61	LI
44123	85C	7.10.63	40016	DERBY	1.9.61	GENERAL
44045	85C	7.10.63	46926	DERBY	15.6.61	GENERAL
73068	85C	9.10.63	21698	DERBY	5.2.63	GENERAL
92157	21A	17.10.63	167000	CREWE	20.11.57	NEW
44263	21A	23.10.63	81000	DERBY	1.12.61	GENERAL
45464	16C	30.10.63	29000	CREWE	15.12.62	INT
92155	21A	15.12.63	175000	CREWE	5.11.57	NEW
44180	21A	23.12.63	74800	DERBY	6.4.60	H INT

Key:

H GEN: Heavy General; INT: Intermediate; LI: Light Intermediate; L CAS: Light Casual; New: Mileage since new.

L.M.S.R. Locomotive Casualty Report System

Methods used for reporting, and dealing with, engine failures

By Harold Rudgard, A.M.I.Mech.E., M.I.Loco.E., M.Inst.T.,
Superintendent of Motive Power, L.M.S.R.

IT is the responsibility of the Chief Operating Manager to supply engines of the correct classification in good mechanical condition to work trains, manned by well-trained and responsible enginemen, at the times they are required. The periods, either mileage or time, at which the various parts of engines are examined and at which boilers are washed out are agreed by the Chief Mechanical Engineer before adoption, and the period at which boilers are examined are laid down by him. The organisation and detailed working in connection with these matters is delegated by the Chief Operating Manager and Chief Mechanical Engineer to a Superintendent of Motive Power.

So that this can be carried out it is essential that there should be a well understood and organised system of examination of the engines at motive power depots. The L.M.S.R. has provided for this by laying down a system, which is adopted at its motive power depots, and which has as its object the maintaining of engines in good mechanical condition in order, as far as possible, to prevent failure of an engine in traffic ; this entails the examination of, and carrying out repairs to, component parts of the engine and boiler on a pre-arranged basis.

Broadly, those examinations carried out on a period basis are to static parts of the locomotive and those on a mileage basis to moving parts. This can be illustrated by mentioning that examinations of injectors, gauge frames, boilers and fireboxes, vacuum-brake apparatus, steam-heating apparatus, etc., are carried out on a periodical basis, whereas examinations to valves and pistons, big and little ends, wheels, tyres and axles, crank axles, etc., are done on a mileage basis.

In all, 64 detailed parts of the engine are examined on a period basis and 41 parts on a mileage basis, and the following list of parts and fittings examined on a pre-arranged basis is given, also with notes for the guidance of District Locomotive Superintendents, to illustrate the comprehensiveness of the examination system.

PARTS EXAMINED

Air valves.
Automatic vacuum or auto steam and vacuum brake.
Axles, crank. Solid or built up.
Axleboxes, engine bogie, truck and coupled wheel.
Axleboxes, tender.
Big ends.
Blast pipe.
Blowdown, continuous, valve and fittings.
Boilers.
Bogie, engine.
Brakes, Westinghouse.
By-pass valve, pipes and connections.
Coal pusher.
Coal bunker, rotary.
Connecting and coupling rods.
Crank pins.
Crossheads, valve spindle.
Cylinders.
Cylinder cocks.
Desanding water apparatus.
Deflector plate, smokebox.
Drawgear.
Ejectors.
Feed trays, top.
Gauge frames and trial taps.
Gauge glasses.
Gauge, boiler-steam pressure.
Injectors, live-steam.

Injectors, exhaust.
Injectors, delivery pipes in boiler.
Indicators, speed.
Joints, pipe.
Joy's valve gear.
Jumper blast pipe top.
Little ends.
Lubricator pipes, flexible, to engine axleboxes.
Lubricator regulator.
Lubricator, mechanical.
Lubricator fountain.
Manifold master valve.
Non-return valves.
Oil pipes.
Pick-up, water apparatus.
Piston rings and heads.
Piston rods.
Piston valves.
Ports, steam and exhaust.
Regulator control gear, vacuum.
Reversing gear.
Sand gun.
Sanding, mechanical, trickle.
Sanding, steam apparatus.
Slide valves.
Smokebox fittings.
Steam-heating apparatus.
Tanks, tender and side.
Valves, steam-pressure safety.
Wheels and tyres

The main examinations causing steam to be out of the boiler are the boiler and firebox examinations, and washing out coincides with these examinations. (Methods used in cooling down and washing out locomotive boilers were referred to in *The Railway Gazette* of May 5, 1940). All other examinations can be carried out during the period engines are stopped for boiler and firebox examination. In carrying out these arrangements, no locomotive should be out of service stopped at sheds for 24 hours, other than for boiler repairs.

If, because of special circumstances, it is considered necessary to carry out any of the examinations more frequently than is provided for in these instructions, or to introduce examinations of other details not mentioned, full particulars are reported to Headquarters for approval. Whenever an engine is stopped for repairs which will necessitate the engine being out of service two days or longer, the District Locomotive Superintendent, Running Shed Foreman, or his representative, after completion of the repairs, and before the engine is put to work, must make a thorough personal examination of the engine. A record is kept by the District Locomotive Superintendent or Running Shed Foreman of every engine thus examined, and is produced when required.

In the case of old type engines of constituent companies examination of details not covered in these instructions, are carried out to the requirements of the Divisional Superintendent, or Operating Manager (Northern Division).

It should be made clear that when examinations of the various parts are undertaken it is laid down that all repairs and adjustments found necessary as a result of the examination are carried out before the engine is put back into traffic. Since the amalgamation of the constituent companies of the L.M.S.R. into one company the total L.M.S.R. engine stock has been reduced gradually from 10,346 to 7,717. It will, therefore, be seen that to enable the Superintendent of Motive Power successfully to fulfil his obligations it has been necessary to ensure that each engine is used to the maximum extent each work-

ing day, and that the total engines out of service each day for repair and maintenance is reduced to a minimum and also that the failures to engines in traffic be eliminated as far as possible.

To achieve these objects the diagramming of engines has been given detailed attention, with the result that the figure of miles worked per engine per day has been very considerably improved. The larger motive power depots have been provided with up-to-date coaling plants, ash-handling plants, etc., to speed up the turn round of engines, and the organisation of the washing out of boilers, examination and repairs to engines, has been overhauled. A serious view always has been taken of engine failures as indicative of either a failure to maintain engines in good mechanical condition by the Superintendent of Motive Power or defects in design, and, so that these two aspects of the same problem can be constantly kept under review, a well thought out and rigidly applied system of reporting failures to engines has been adopted by the L.M.S.R., which is called a casualty system.

The casualty system now in use is an extension of the ideas incorporated in the original system of reporting casualties on the old Midland Railway and the author of this article was closely associated with the late Sir Cecil Paget in bringing into being this method. The casualty system also brings to light any failure on the part of the motive power staff to carry out the instructions governing the examination and repair to locomotives, or cases of bad workmanship resulting in failure of an engine, but this, although necessary, is considered as of secondary importance to the main object of the casualty system already mentioned ; the primary object is to ascertain the real cause of the casualty, with a view to eliminating similar cases.

In the first years after amalgamation, the casualty system was operated from Headquarters, but a change has been made and the present instructions provide for all casualties being reported by District Locomotive Superintendents to their respective Divisional Superintendents at Crewe, Derby and Manchester, and to the Operating Manager, Glasgow. The Divisional Superintendents and Operating Manager, Glasgow, submit four-weekly statements of casualties under the following headings to the Superintendent of Motive Power :—

1. Summary of debitable casualties.
   A debitable casualty is defined as a mechanical casualty where three or more minutes have been lost.
2. Analysis of casualties—mechanical.
3. Analysis of casualties—other than mechanical.

The system of reporting casualties thus provides for detailed examination of the individual casualty forms in the Divisional Superintendents' office, and the following records are kept at Headquarters :—

1. Four-weekly summary and aggregate summary of debitable casualties.
2. Four-weekly summary and aggregate summary of mechanical casualties.
3. Four-weekly summary and aggregate of "other than mechanical" casualties.
4. Four-weekly and aggregate statements of miles run per debitable casualty.
5. Four-weekly and aggregate statements of miles run per mechanical casualty.

These records enable the Superintendent of Motive Power to observe the general trend of casualties under any particular classification and if necessary the Superintendent of Motive Power, calls for detailed information from the Divisional Superintendents or Operating Manager.

Dealing with casualties is looked on by a

District Locomotive Superintendent as of primary importance and for purposes of identification the casualty report forms on each division were printed in distinctive colours, namely :—

Northern division	..	Pink
West division	..	Blue
Central division	..	Green
Midland division	..	Yellow

and, therefore, not only did the form draw attention to itself as something requiring special attention, but the use of a definite colour for each division enabled the District Locomotive Superintendent to realise immediately whether he was dealing with a casualty affecting his own division or one to an engine allocated to another division. Coloured forms are not now in general use

delay is in no way due to any defect in the engine :—

Mismanagement	Slipping
Shortage of water	Shortage of coal
Overloading	Priming
Weather conditions	

Category IV

Will include all casualties where time has been lost due to inferior coal.

Category V

Will include all cases of steam heating irregularities irrespective of whether time has been lost thereby or not.

A driver must specially report all cases of time lost by locomotive or engine failure during the working of his train, when he signs off duty whether he finishes at home shed or another shed :—

The report is to be handed to the running shift foreman who must satisfy himself that all the necessary information is given to enable the case to be properly reported. Brief particulars of time lost with all important passenger and freight trains and serious engine failures must be immediately telegraphed to Superintendent of Motive Power, Watford, and appropriate Divisional Superintendent. An examination of the engine must be made immediately (where possible) with a view to giving the correct cause in the telegrams.

Fractures

In all cases of fracture of any part of the engine motion or of an axle, wheel, tyre, or drawgear, whether there is any delay or not, a telegram must be sent immediately to the Superintendent of Motive Power, Watford H.Q., the appropriate Divisional Superintendent of Operation, the Chief Mechanical Engineer, Derby, and the Mechanical Engineer or Locomotive Works Superintendent, stating where the broken parts may be inspected.

Category I

A. An engine casualty report form giving as many particulars as possible must be made out at the shed at which the driver signs off. Against items on the report which cannot be completed, the words " To follow " or the name of the shed responsible for the information must be inserted. A note must also be made on the engine casualty report form that a copy has been sent to the shed or sheds concerned. Below is a specimen copy of the form used to complete an engine casualty report.

Copies of the engine casualty report

Specimen of driver's report

because of the difficulties of wartime conditions.

Instructions for the working of the casualty system, which include the classification and coding of casualties, are given below.

Engine casualties are dealt with under five categories :

Definition of Engine Casualties

Category I

1. During the working of a train any defect developed in the engine or tender whereby the efficiency is impaired and time is therefore lost.

2. The failure of an engine, after being fired, to work its train to time through any defect which should have been found by the examiner or driver or was due to a concealed defect.

3. Fusible plugs, irrespective of whether time has been lost or not.

(Items 1 and 2 of the above are reportable to the Board of Directors of the company).

Category II

Will include casualties under the following headings where no time has been lost :—

1. Hot axle bearings which necessitate the removal of wheels, and hot tender axle bearings which necessitate removal of wheels or bearings.

2. Hot big or little ends.

3. Engines unable to return to owning depot with train booked through insufficient time to carry out the repairs necessary.

Category III

Will include casualties under the following headings where time has been lost, but it must be emphasised that particular care must be exercised to ensure that the

Engine casualty report form, to be completed at " signing off " shed

| LMS | WEEKLY STATEMENT OF ENGINE CASUALTIES (MECHANICAL) |

MOTIVE POWER DEPOT_____ WEEK ENDED_____19__

NUMBER OF ENGINES ALLOCATED Passenger_____
Freight_____

| DATE | ENGINE NUMBER | CLASS | SHED | TRAIN WORKED OR TRAIN WHICH SHOULD HAVE BEEN WORKED | | DELAY | NATURE OF CASUALTY | REMARKS | Ref. No. |
				TRAIN P = PASSENGER F = FREIGHT	FROM	TO				

| LMS | WEEKLY STATEMENT OF ENGINE CASUALTIES | LMS |

(OTHER THAN MECHANICAL)

| DATE | ENGINE NUMBER | CLASS | SHED | TRAIN WORKED OR TRAIN WHICH SHOULD HAVE BEEN WORKED | | DELAY | NATURE OF CASUALTY | REMARKS | Ref. No. |
				TRAIN P = PASSENGER F = FREIGHT	FROM	TO				

Specimen of headings on front and back of weekly statement of engine casualties

form are to be forwarded without delay as follow :—

(1) One copy to the appropriate Divisional Superintendent of Operation (two in cases of fractures).

(2) One copy to shed where disabled engine was left.

(3) One copy to shed at which disabled engine is allocated.

(4) One copy to shed at which driver is stationed.

B. The District Locomotive Superintendent or Running Shed Foreman of the shed where the disabled engine is left must have an examination made to ascertain the cause of the casualty, and after receipt of the copy of the engine casualty report form must send in letter form to the appropriate Divisional Superintendent of Operation the result of the examination and the cause of the casualty.

If a copy of the engine casualty report form is not received by this District Locomotive Superintendent or Running Shed Foreman within three days of the casualty, he must make out an engine casualty report form and forward copies in accordance with (1), (3), and (4) above, also a copy to the shed at which the driver signed off duty. In such cases the District Locomotive Superintendent or Running Shed Foreman of the shed at which the driver signed off duty must send to the appropriate Divisional Superintendent of Operation an explanation why the engine casualty report form was not initiated by him.

C. The District Locomotive Superintendent or Running Shed Foreman of the shed to which the disabled engine belongs, on receipt of the engine casualty report form, must forward to the appropriate Divisional Superintendent of Operation a letter giving all the particulars for which he is responsible.

D. If a casualty occurs in the district to which the engine and driver belong, the engine casualty report form, if possible, should be fully completed on both sides in the first instance. If this is not possible, the words " To follow " should be inserted opposite " cause of casualty " on the front side of the Report and the reverse side should be left blank. The reason why the case cannot be completed should be included under " Remarks."

E. In a case where the engine casualty report form has not been completed in the first instance, the report and all letters connected therewith will be sent for conclusion to the shed considered responsible for the casualty. The reverse side of the engine casualty report form must then be completed fully by the District Locomotive Superintendent or Run-

L M S Shed_____ Date_____19__

ENGINE CASUALTY REPORT

Shed_____ Engine No._____ Class_____ Date_____19__

Driver_____ Fireman_____ Stationed at_____

Shed_____ Asst. Eng. No._____ Class_____ from_____ to_____

Driver_____ Fireman_____ Stationed at_____

Train_____ m._____ from _____

Delay_____Hrs._____Mins. At Between_____ and_____ Load { Regulation_____ Actual_____

Nature of casualty_____

Cause of casualty _____

Part at fault_____ | Last Heavy Repair in C. M. E. Shops.

Date last renewed_____ | Date_____ Shop_____

Date last examined (Daily)_____ By whom_____ Shed_____

„ „ „ (Standard)_____ By whom_____ Shed_____

Has this part been reported on } By whom_____ Shed_____
any of the six previous trips } Date _____

If so state:—What was done_____

By whom_____ Date_____ Shed_____

Particulars of the six previous trips :—

	Date	Driver	Mileage	From	To
1					
2					
3					
4					
5					
6					

Engine { gave train up / exchanged } at _____ { to / for } Engine No._____   Time Casualty Occurred

Dist. Loco. Supt.'s or Running Shed Foreman's Remarks and Summary of Driver's Report:—

Divisional Superintendent of_____ Signed _____
District Loco. Supt. or Running Shed Foreman.

[over

Front of engine casualty report form

ning Shed Foreman (special care being taken to give under heading "Cause" particulars relevant only to the cause of the casualty) and returned as quickly as possible to the appropriate Divisional Superintendent of Operation.

F. Time lost and regained

In cases of time lost during one part of the journey being regained later on, this should be clearly indicated and the points given between which the time was regained.

G. Written reports from shed or repairing staff

Written reports from the shed or repairing staff are not to be obtained in connection with engine casualties. Drivers' reports or reports from foremen are not to be forwarded with engine casualty report forms, but must be summarised by the District Locomotive Superintendent or Running Shed Foreman and inserted together with his remarks in the space provided on the front of the engine casualty report form, or in the letter, when one is sent in accordance with the above instructions.

H. Casualties to engines working double-headed trains

In order to make it clear on the engine casualty report form when the assisting engine has failed, this engine number must be underlined with red ink.

J. Particulars of last general intermediate or service repair, last standard examination

Correct particulars as to date out of Chief Mechanical Engineer's Shops after general intermediate or service repair, and mileage since each class of repair must be given. In addition to giving the date of last standard examination, show in cases of examination on the mileage basis, the mileage run from the date of such examination to date of casualty, and mileage since last "X" examination and repair.

K. Fusible plugs fused

All cases of the above are to be reported on form E.R.O. 47987, shown on previous page, irrespective of whether time has been lost thereby or not. Disciplinary action must not be taken in these cases, but a recommendation sent to the Divisional Superintendent. Details of mileages for the last three washouts to be given, also any priming reported on the last six trips.

I. All casualties under Category I must be shown on the Weekly Casualty Statement E.R.O. 53921 for the district to which the engine is allocated in accordance with the instruction contained under the heading "Weekly Engine Casualty Statement E.R.O. 53921." A specimen copy of the weekly statement of engine casualties is reproduced on the previous page.

Category II

Such cases are not to be reported on casualty report per E.R.O.47987 but must be shown on the weekly statement E.R.O.53921 of the District to which the engine is allocated. In cases where the engine fails away from the home station due to heated axle, big and little end bearings (where time has not been lost), the non-debitable casualty report form E.R.O.53920 must be made out by the District where the engine is stopped, giving full particulars of the result of examination, and the apparent cause. Alongside is reproduced a specimen of the non-debitable engine casualty report.

One copy of this form must be forwarded within three days to the Divisional Superintendent of Operation to whom the engine is allocated, and one copy to the District to which the engine is allocated. If it is not possible to make the examination within three days a report must be sent, but the words "particulars to follow" inserted, and as soon as examination is made the results should be sent in letter form to the Divisional Superintendent of Operation, and the District Locomotive Superintendent to whom the engine is allocated, but it must be strongly emphasised that every effort must be made to carry out examination of heated bearings as soon as possible.

The District Locomotive Superintendent to whom the engine is allocated, after receiving the report, will make full investigations, and deal with the case, recording it on the weekly casualty statement E.R.O.53921, giving briefly the result of his inquiry and action taken, and also details of the oil supplied.

In cases where an engine is unable to return to the owning Depot on a booked train, a non-debitable casualty form E.R.O.53920, must be made out by the District Locomotive Superintendent where the failure occurred, giving full particulars of the nature and cause of the casualty. One copy must be sent within 24 hours to the appropriate Divisional Superintendent of Operation and one copy to the District Locomotive Superintendent to whom the engine is allocated, who must fully investigate and deal with the case, recording his conclusion and action taken on the weekly casualty statement.

Category III

All cases of loss of time due to any of the following causes, must be recorded on the back of the weekly casualty statement, E.R.O.53921, by the District where the driver belongs. The driver's reports should not be sent to Divisional Superintendent of Operation unless specially asked for.

Mismanagement
Shortage of water
Overloading
Weather conditions
Slipping
Shortage of coal

In cases where the driver books off at a foreign shed, the District Locomotive Superintendent should forward the driver's report form to the District where the driver is stationed, and insert any remarks he may wish to make at the foot of the report.

Priming

Cases of boiler priming are to be reported on the back of weekly casualty statement, E.R.O.53921, by the District where the engine is allocated. In cases where the driver books off at a foreign station, the driver's report form must be suitably endorsed with brief details as to the condition of water and any points relevant to the case, and sent to the District Locomotive Superintendent to whom the engine is allocated.

Inferior Coal

Category IV

When it is considered that time has been lost because of inferior quality of coal, a coal report form, E.R.O.23299, must be initiated, but it is essential that the District Locomotive Superintendent shall satisfy himself that the quality of the coal is responsible for the delay. When an engine has taken coal at a mechanical coaling plant the particulars of coal in the plant at the time the engine took coal and this, of course, may be a mixture of coal from more than one colliery.

A copy of this form should be sent to:—

(1) Divisional Superintendent of Operation.

(2) District Locomotive Superintendent where engine is allocated.

(3) District Locomotive Superintendent where engine last coaled.

(4) District Locomotive Superintendent where men are stationed.

(5) Coal Office, Derby (English Divisions).

(6) Coal Office, Glasgow (Northern Division).

These cases are to be recorded on the back of weekly casualty statement of the District where engine is allocated.

Category V

Cases of steam heating irregularities should be reported on the back of the weekly casualty statement, E.R.O.53921, of the District to which the engine is allocated irrespective of whether time has been lost or not. In cases where the driver books off at a foreign station, the driver's report form must be suitably endorsed with points under item number 16 (in notes for guidance in dealing with certain engine casualties) and sent to the District Locomotive Superintendent to whom the engine is allocated.

LMS SHED_____ DATE_____

NON DEBITABLE ENGINE CASUALTY REPORT

Your Engine_____ Class_____

failed after working_____

on _____through the following defects :—

Non-debitable engine casualty report form

COMMON CAUSES FOR ENGINE FAILURES

(extract from The Steam Locomotive in Traffic by E A Phillipson, Locomotive Publishing Co.)

SOME COMMON CAUSES OF SPECIFIC FAILURES.

Hot boxes.
Insufficient or unsuitable oil supply: for example, excessive lowering of viscosity with rise of atmospheric temperature. Dirty or unsuitable trimmings. Pads dirty and/or glazed, or not tip to journal owing to weak or broken pad springs. Sand, grit or ash on journal. Water in oil reservoir or keep; original sources are rain, leakage from tender or side tank, carelessness whilst taking water or during washing out, overflow from Injector, joints in water pipes leaking, or condensate from blows at joints in steam pipes. Blockage or fracture of oil pipes. Slack joint in oil pipe. Oil pipe from framing not registering with intake on box. Keep broken, lost or rubbing on journal. Inserted brass broken or loose. Unsatisfactory fitting of brass to journal: not properly bedded down on crown and tight at sides, brass bearing hard on radius (where collars are used), box tight in horns, oilways cut into radii, oilways cut in whitemetal and not in brass. Bearing springs weak or broken, spring pins broken or weight distribution faulty owing to other causes. Opposite axleboxes out of alignment: for instance, wedges unequally adjusted. Whitemetal defective. Failure of mechanical lubricator, such as pump not working, pipe union slacked back, check valve defective, lubricator loose on bracket or feed incorrectly set. Reservoirs in crown of box inaccessible for replenishment and lubrication, owing to engine riding low in frames. Engine overworked. Engine prematurely rostered on arduous turn subsequent to re-fitting of box. Effect of previous derailment.

Some types of axlebox have an insufficient capacity to disperse by radiation the heat normally generated or are more prone to distortion, and therefore more liable to heat, than others. The disposition of oilways is also a deciding factor; care must be taken, when designing, with their location in the case of coupled boxes, and they should be eliminated as far as possible for carrying boxes, for which bottom feed lubrication is emphatically preferable.

Hot big ends.
(It will be appreciated that many of the causes of heating in axleboxes apply also to big ends and other bearings). Plug trimmings too tight. Plug or wire trimmings too short, causing oil to be delivered between strap and brasses. Abrasive foreign matter on journal. Cotter slacking back owing to loose or broken studs. Brasses loose in strap or broken. Crank pin worn oval. Insufficient lateral play allowed. Small ends out of alignment. With inside big ends, dowels working out of built-up crank axle webs. Excessively short connecting rods. Excessive compression of engine when notched up. Excessive rigidity of engine as a vehicle.

Fractured or distorted big end straps.
Brasses loose. Water trapped in cylinders as, for instance, when priming. Bolts not properly fitted.

Fractured or shouldered big end bolts.
Cotters not properly fitted. Brasses loose.

Overheated slidebars.
Lubrication interrupted by glands blowing. Sand or grit on bars. Slidebar bolts loose or slidebar out of alignment through any other cause.

Damaged piston rod or sheared crosshead cotter.
Obstruction in cylinder. Water trapped in cylinder as, for instance when priming. Piston rod insufficiently lubricated, especially when engine is coasting. Slide blocks seized.

Fractured piston head.
Too tightly fitted on rod. Water trapped in cylinder, due to priming or other cause. Portions of valve or piston rings, or other obstructions, in cylinder.

Piston glands blowing.
Insufficient compression on packing spring. Rod dropping owing to wear. Rod either bent, worn taper or not truly circular. Packing not properly bedded to rod.

Cylinder pressure relief valves blowing or sticking.
Grit on seating. Spring made up with carbon or other foreign matter.

Cracked inside cylinders.
Worn thin on top by action of smokebox ash.

Valves seizing.
Rocking shaft bent or loose. Failure of mechanical lubricator (as with hot boxes). Excessive lubrication, causing carbon to deposit under rings.

Defective reversing gear.
Trigger fractured (fatigue hardened). Reversing rod jammed when engine negotiating curve. Oil leaking from cataract cylinder (power reverse). Excessive valve friction. Valve spindle seized or too tight.

Eccentric straps seized or fractured.
Valve, motion or strap itself insufficiently lubricated, due in some cases to main bearing surface being arranged on the minor, as opposed to the major diameter of the sheave. Strap bolts not cottered, or otherwise improperly secured. Abrasive foreign matter on bearing surface. Excessive carbonisation of valve.

Hornblock wedges dropping.
Vibration. Wedge bolts not securely locked.

Fractured laminated bearing springs.

Thick back plates. Fastenings of unsatisfactory design. Absence of india-rubber auxiliaries. Insufficient flexibility of engine. Poor condition of permanent way.

Bent or fractured coupling rods.

Excessive slipping. Uneven distribution of sand. Material fatigued or unsuitable. Insufficient flexibility in the case of multi-coupled engines. Rod collars not properly secured.

Defective sanding gear.

Supply of sand exhausted. Pipe not set to rail or not discharging directly under wheel. Sand blown off rail by wind. Steam pipes burst by frost or other cause. Pebbles in valve or trap. Wet sand in box, valve or trap (water may be thrown up from the wheels, enter through sand box bolt holes, or through the box lid if the latter is too shallow or not tight; sanding steam valves may blow through and condense). Operating gear jammed when overtravelled, and pins brought into line with fulcrum in consequence. Operating rods whipping.

Defective brake rigging.

Brake stretcher ends, pull rods, brake hanger brackets, or bracket studs fracturing owing either to wear, defective material or unsatisfactory workmanship. Threads stripping on pull rods or turnscrews.

Defective steam brake.

Cylinder insufficiently lubricated. Piston rings faulty. Cylinder cold. Combination valve not properly adjusted. Release spring weak or not properly adjusted.

Defective vacuum brake.

Rolling ring twisted. Cylinder oily, smooth or rusty. Cylinder trunnions not lubricated. Piston rod not screwed home. Ejector cones badly worn or scored, loose, or made up with scale. Drip valve sticking. Excessive vacuum generated by engine previously working train. Joints not airtight.

Defective Westinghouse brake.

Pump reversing spindle bent or fractured. Triple valve dirty. Driver's valve dirty. Governor faulty or incorrectly set.

Regulator stiff to operate.

Gland packed too tightly. Valve (slide type only) insufficiently lubricated.

Defective hydrostatic lubricator.

Feed nozzles or chokes made up.

Fractured draw gear.

Material fatigued or excessively worn. Mismanagement of engine. Varying brake power on individual vehicles of train. Vehicles slack coupled.

Blower ineffective.

Blower holes either enlarged by attrition or made up. Steam pipe joints slacking back. Steam pipe bursting owing to corrosion.

Injectors not working.

Feed pipes, or joints, drawing air. Flexible feed pipes burst or leaking. Tank sieve made up or faulty. Cones encrusted with scale or obstructed by foreign matter. Cones broken, distorted or worn, excentric or loose. (Distortion may be caused by cleaning the cones with a file or scraper). Delivery or overflow pipes made up. Engine short of steam. Water supply in tank too hot. Steam supply to injectors wet. Injector overheated, due to steam cock blowing, or clack blowing back by reason of wear or foreign matter on seating. In the case of combination injectors, there may be excessive exposure to heat from the boiler backplate. Where non-automatically controlled exhaust steam injectors are concerned, overheating may be caused by the omission on the part of the enginemen to close the exhaust steam wing valve when the feed is shut off.

Fractured piping.

Material fatigue hardened owing to vibration. Insufficient allowance made for expansion. Faulty brazing of collars.

Steam pipe joints blowing.

Nuts slack on studs, due in some cases to inaccessibility. Leakage and perishing of jointing between two adjacent stud holes, the pitch of which may be too fine from this point of view. Insufficient number of studs provided, encouraging uneven distribution of pressure when hitting. Jointing material too thick or unsuitable. Undetected grooving (in the case of metal to metal joints)

Vibration.

Excessive expansion.

Drop grates seizing.

Trunnions blocked with ash. Operating gear distorted.

Uncontrolled movement of drop grates.

Lock failing or not used, in conjunction with wear of operating gear.

Collapsed brick arch.

Unsatisfactory building. Defective material. Engine steamed prematurely after building of arch. Misuse of fireirons.

Fusible plugs blowing.

Filling not homogeneous, or deteriorated. Formation of scale on top of plug. Insufficient thread contact with crown plate. Wear of threads.

Tubes leaking and bursting.

Use of corrosive or heavy scale forming waters. Omission to remove scale from boiler completely when cleaning. Excessive stretching of tubes in shops or imperfect welding when piecing up. Improper use of tube expanders, particularly if of the three-roller type (expanders with five rollers are preferable in every respect). Excessive use of blower, causing tube ends to burn. Firehole door left open more than necessary. Opening of smokebox door subsequent to cleaning of fire, or for excessive periods. Dampers left open subsequent to throwing out of fire. Excessive watering of coal. Misuse of liquid fuel burners, the heat from which is essentially intense and localised.

Engine priming.

Poor quality of water. Excessive mileage run since previous wash-out. Unsatisfactory washing out, such as at insufficient pressure or rodding not performed thoroughly. Water level in boiler too high. Engine worked too hard. Mixture of treated with raw feed waters. Feed water treatment not appropriate.

Engine refusing to start.
Brakes not fully released. Engine standing on dead centre. Engine overloaded. Sanding gear inoperative. Valves or pistons blowing. Weight transferred from coupled to carrying wheels, or not uniformly distributed over the former.

Engine not steaming freely.
Fuel of unsatisfactory quality or stale. Engine in run down condition or overloaded. Engine mismanaged, for instance, by being worked unnecessarily hard, firing too heavily or with coal of excessive size, or firehole door left open to an abnormal extent. Men inexperienced with type of engine or class of work concerned. Defective firebars or dampers. Tubes or stays leaking. Tubes dirty or made up. Brick arch to incorrect dimensions or otherwise defective. Valves fractured, blowing or requiring to be reset. Pistons blowing. Smokebox drawing air. Joints blowing or pipes burst (usually owing to corrosion or burning) in the smokebox. Blast pipe out of alignment with the chimney (if in the transverse plane, this defect may be observed from the cab when the engine is running with the regulator open). Blast pipe orifice not at correct height in smokebox, or unevenly carbonised. Cowl out of alignment. Blower foul of blast pipe orifice. Superheater element burst. Internal delivery pipe defective, and having an adverse effect in consequence on the movement of the convection currents in the boiler.

Jobs to be done!

These pictures, taken by members of the Chief Civil Engineer's Photographic Section, show three essential tasks being carried out at Gloucester Barnwood depot in 1962: tube cleaning; firebox examination; and welding in the Blacksmith's shop.